DESIGNING WITH LIGHT

Public Places

Lighting solutions for exhibitions, museums and historic spaces

DESIGNING WITH LIGHT

Public Places

Lighting solutions for exhibitions,
museums and historic spaces

JANET TURNER

Series Editor: Conway Lloyd Morgan

A RotoVision Book
Published and distributed by RotoVision SA
Rue du Bugon 7
CH-1299 Crans-Près-Céligny
Switzerland

RotoVision SA, Sales & Production Office
Sheridan House, 112/116A Western Road, Hove
East Sussex BN3 1DD, UK
Tel: +44 (0)1273 72 72 68
Fax: +44 (0)1273 72 72 69
E-mail: sales@rotovision.com

Distributed to the trade in the United States by
Watson-Guptill Publications
1515 Broadway
New York, NY 10036

ISBN 2-88046-333-5

10 9 8 7 6 5 4 3 2 1

Book design by Lovegrove Associates

Production and separations in Singapore by
ProVision Pte. Ltd.

Tel: +65 334 7720
Fax: +65 334 7721

Acknowledgements

The author, editor and publisher wish to thank the following for their help, advice and encouragement in the creation of this book: Patricia Aithie, Max von Barnholt, Martin Caiger-Smith, Steven Dean, Andrew Dempsey, Craig Downie, Terry Farrell, Ray French, Barry Gasson, Matthew Goer, Piers Gough, John Hobbs, Alan Irvine, John Johnson, Jan & Amanda Kaplitsky, Alexander Kovacs, Charlotte Kruk, Bob Lock, Andrew Logan, Benjamin Magnússon, John Miescher, Jean Nouvel, Terry Pawson, Tom Porter, Julian Powell-Tuck, Philippe Ruault, Alan Stanton, Paul Thompson, Jane Wess, Simon White, Keith Williams and Peter Wilson, as well as the following companies and organisations whose work appears here: John Dangerfield & Associates, Electricité de France (EDF), Erco, Future Systems, Lightscape Inc., Pawson Williams Architects, Philips, Stanton Williams Architects, Studio Downie and in particular colleagues and friends at Concord Sylvania, SLI worldwide and the Chicago Miniature Lamp Co.

Special thanks go also to John Hobbs Ltd. and Purves & Purves for allowing us to photograph and feature works from their respective showrooms, and to the Hayward Gallery and Anish Kapoor for permission to show the latter's piece, _Double Mirror_, 1997–1998, from an exhibition of his work at the Hayward Gallery, London in May 1998.

Very special thanks go also to Keith Lovegrove for his fine work on the design, and to Natalia Price-Cabrera for her editorial vigilance.

Janet Turner
London, August 1998

The photographs reproduced in this book are copyright. We would like to thank the following photographers for permission to reproduce their work: Patricia Aithie, Richard Bryant/Arcaid, Peter Cook/VIEW Ltd., Richard Davies, Dennis Gilbert, Kita, John Edward Linden, Claude Pauquet, Andrew Putler, Paul Raesied, Philippe Ruault, Malcolm Robertson, Timothy Soar, Steve Theodoras, Nigel Young and Nathan Willock.

Section One:

BEGINNING ON THE BANK   8

Section Two:

LIGHTING BASICS   18
Light   22
Perception   23
Colour   24
Sources of Light   26
Colour of Light   28
Colour Rendition   29
Artificial Light   30
Lamp Types   32
Summary Guide to Lamp Types   35
Fitting Types   36
Output, Efficiency and Cost   38
Light Distribution   40
Intensity and Diffusion   41
Light and Visual Effect   43
Lighting and Conservation   44
Lighting as a System   50

Section Three:

FACING THE FAÇADE   52
Louvre Museum   58
Goodwood Visitor's Centre   64

On the illustrations in this book the symbols
below describe the type of directional lighting
used, with the beam widths *(facing)* as
appropriate. These symbols relate to the main
lighting system in the relevant illustration.

Direction of Light

 Downlighting   Uplighting   Sidelighting   Spotlighting   Multi-directional lighting

Section Four:

RESOLVING THE INTERIOR	70
LIGHTING ON THE LINE	72
The Atlantis Gallery	80
Duveen Wing	84
Mondrian Exhibition	90
WORKING IN THE ROUND	92
Bodleian Library	96
Earth Galleries	102
The National Gallery of Modern Art	110
Welsh Contemporary Crafts Exhibition	114
George III Gallery	118
Burrell Collection	124

Section Five:

COMPLETING THE TASK	130
Birmingham School of Arts	132
Benetton Sport '90	138
Carrickfergus Heritage Centre	142
Cartier Foundation	146

Section Six:

CONCLUSION	152
GLOSSARY	156
FURTHER INFORMATION	158
INDEX	159

Beam Widths

 Narrow beam
 Medium beam
 Broad beam

The Hayward gallery on the South Bank of the Thames in London was built in the early 1960s. It was intended to house temporary exhibitions of works of art of all kinds, especially contemporary paintings. Its exterior is a Brutalist mass of shuttered concrete, and the same surfacing is used on ceilings and staircases within the complex. Within, the walls are finished in plain white with concrete ceilings and plain coloured floors.

The interior spaces can be modified in plan by the use of partitions, and by inserting false ceilings to change the proportions of exhibition spaces to suit the work on view. This flexibility is increasingly useful as the range of work exhibited evolves and changes.

The gallery comprises a linked series of large exhibition rooms on three levels. It is a very uncompromising space, at first sight. While it forms a splendid backdrop to abstract expressionist or colour field paintings, or for large-scale sculptures, it is a less easy setting for smaller scale, figurative or more intimate work, where the solid volumes and stone colours can prove daunting. Over the years Concord Lighting has been involved in lighting many different exhibitions at the Hayward, and in doing so has experienced a whole range of challenges. Sculptures by Rodin, paintings by Rothko, native works of art from South America, icons from Greece, posters from Soviet Russia, architecture from Edwardian England, each of these (and many others) required different lighting solutions from the same tracked system.

Different solutions, but the same approach in all cases: the crucial factor has been to work closely with the exhibition designer from as early a stage as possible. It is important to understand the general concept behind the exhibition, and get a feeling for how the organiser and designer want to see the work displayed; whether it be in a formal, or informal, manner. How is the work to progress through the different exhibition spaces, and what are the aims of the exhibition? This familiarisation process runs right through all the stages of the job, and even if the impression is at the start somewhat vague and unformed, still getting this understanding right is a good way of making sure that the final, technical solution is in tune with the aims of the exhibition organiser.

The next stage in the design work is to make a lighting audit. This has firstly to look at the existing lighting systems both for fittings and lamps and for lighting positions, and at the circuitry and alternative switching possibilities that are available. If there is any reasonable budget for lighting, it may be possible to add to the existing material, or it may be a question of having to make what is already to hand do the job. The second, but equally important part of the audit consists of analysing the works of art that are to be displayed, to ascertain whether any of them require special lighting considerations from the conservation perspective (this question is discussed in more detail at the end of the next chapter) and whether in addition, there are particular works which are the high points of the exhibition and so will need special lighting consideration. It is an essential element of good practice that any lighting solution follow the technical advice for light levels given to the designer from a conservation standpoint. All works, with the exception of unpainted metal and stone, suffer in time from the effects of light and so it is a matter of duty to ensure that they receive the minimum lighting exposure so that their beauty and value endure as long as possible. This is a particularly critical area where paper, textiles and natural materials are concerned. Most museums and galleries today have lighting systems that meet conservation standards, but other exhibition spaces (for example, trade fair areas) may not do so.

Traditional Chinese paintings on paper scrolls displayed in cases *(above, facing)*, or contemporary art from the 20th century *(below, facing)* or even Lutyens' model for Liverpool's Roman Catholic cathedral *(above)*: all are lit from the same tracked system.

With a clear idea of the lighting system and the conservation requirements, and a growing understanding of the purpose and intentions of the exhibition, the lighting designer should then move on to looking at lighting solutions room by room. Here the intended visitor itinerary is important. If there are changes in light levels between different areas, it is a courtesy to allow the visitor's eyes time to become adjusted to the new light level. Unless a particularly dramatic effect is looked for, changes in light level should be as imperceptible as possible. How this can be executed in practice can be seen by comparing two very different exhibitions, both shown in the same space.

The 'Twilight of the Tsars' was an exhibition on the fine and decorative arts in Russia at the beginning of the 20th century – a period often overlooked in favour of the later developments of agitprop and constructivist art after the Revolution. But Russian art nouveau, with its special fusion of Western ideas and traditional folk art was flourishing, and symbolist painters such as Nesterov were exploring the cultural contradictions of the period. In the decorative arts, the early genius of designers such as Benois and Bakst was already apparent.

The keynote for the lighting design is there in the title: the individual works – portraits and crucifixes, icons and costumes were highlighted against the pale walls with framing heads on individual spotlights, and the rest of the space left dim. This solution worked well for three reasons: it put plenty of light on to the objects (which were often in ornate materials with detailed workmanship), it reflected the themes of termination and change implicit in the exhibition, and it matched the lighting levels found at the time, for example in Russian churches such as St Basil's cathedral in Moscow.

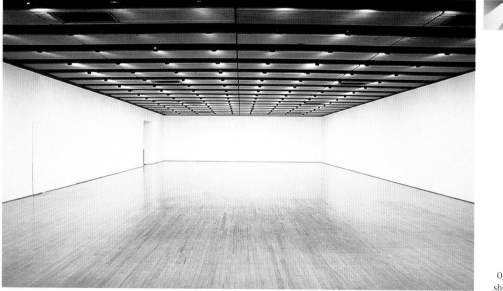

One of the main gallery spaces at the Hayward, showing the lighting system *(left)*.

The theme of twilight and isolation is maintained in
the continuing display and in cases *(above)*.

Framing heads highlight individual objects at the
entrance to the 'Twilight of the Tsars' exhibition *(right)*.

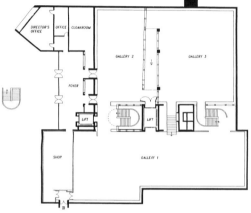

Plans of the main exhibition floors at the Hayward Gallery *(top)*.

At night a programmed neon sculpture highlights the position of the Hayward Gallery within the South Bank *(left)*. Exterior of the gallery *(above)*.

Installation for the 'Gravity and Grace' exhibition, lit by John Johnson *(facing)*.

'Gravity and Grace' was a completely different proposition: an exhibition of abstract sculpture from 1965–1975, a seminal decade in 20th-century art. The works exhibited, by artists such as Joseph Beuys, Barry Flanagan, Richard Long and Eva Hesse among others, sometimes incorporated light sources. While the work in the Russian exhibition had some stylistic continuity, here the themes were interpreted in radically divergent ways by the individual artists.

Exhibits could include black, strobe and neon lights, or lettuces between blocks of granite. One piece featured a live parrot on a perch. When first exhibited, much of the work excited controversy: seen anew at the Hayward Gallery in 1992 after a gap of two decades, the exhibition provided an important opportunity to re-evaluate both individual artists and the progress of sculpture. The key therefore was to plan for an overall level of illumination that would allow the visitors to compare works, together with a system of partitions that would separate works avoiding visual clashes. The solution also had to allow for lower light levels where appropriate for the lighting inherent in the work to play its role.

The lighting designer is, like the exhibition designer, an intermediary between the visitor and the work, both bringing the visitor closer to the work on show, and enabling the work to communicate most appropriately with the visitor. Putting this into practice is one of the most exciting aspects of lighting design.

The versatility of the Hayward Gallery is shown in these two images. In one case *(above, left)*, two emblems from different cultures (Nazi Germany and Soviet Russia) are spotlit independently to heighten the difference: in the other *(above, right)*, an even wash of colour, lights both Soviet paintings and sculpture. Both examples were shown at the 'Art & Power' exhibition, created for The Council of Europe by Andrew Dempsey.

Normally a pure white background makes colour perception clearer but dulls modelling: here at the Yves Klein exhibition the aim was to convey the intense colours of Klein blue, rather than express the sculptural forms (above). When pure white forms need to be seen in detail, a more rarified lighting is called for, as at an exhibition of Mexican art (right).

'I was so persecuted with discussions arising out of my theory of light, that I blamed my own impudence for parting with so substantial a blessing to run after a shadow.' The words are Sir Isaac Newton's, and show just how problematical the understanding of light, perception and colour has been in history. For light is what enables us to see, and what has always enabled humankind to see.

How this process works has, over the centuries, attracted the interest of thinkers such as Aristotle and Goethe, and of scientists such as Newton, Young, Helmholtz and Hering. The nature of light is one of the key questions underlying modern physics in the work of Einstein and his successors, and the representation of light has fascinated artists from Constable and Turner, to Rothko and Vasarely. For while light and our experience of light through seeing are physical phenomena that can be analysed, they are also real experiences which we all share.

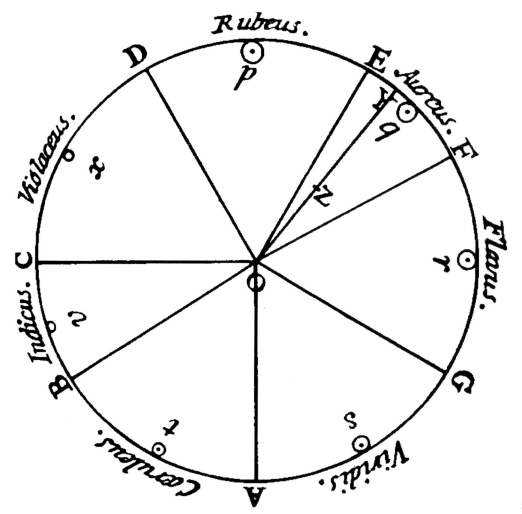

A diagram from Newton's *Opticks* showing his theory of colour *(left)*: a continuous movement from red *(Rubeus)* to orange *(Aureus)*, yellow *(Flavus)*, green *(Viridis)*, blue *(Cæruleus)*, purple *(Indicus)*, to violet *(Violaceus)* and back to red again *(left)*.

A portrait of Sir Isaac Newton by John Vanderbank *(facing)*.

Sir Isaac Newton by John Vanderbank (1694?-1739). By courtesy of The National Portrait Gallery, London.

Light

In technical terms, light is a form of electro-magnetic radiation, which is visible to the human eye in a narrow band between 400 and 800 nanometers in wavelength. (Some animals and birds have a slightly different spectrum of vision.) Shorter wavelengths include infrared light, longer ones ultraviolet light. Within the visible spectrum we distinguish a range of colours from red (the shortest wavelength) to blue. And although there is general agreement as to the wavelengths of specific colours, our own experience of colour is in fact subjective.

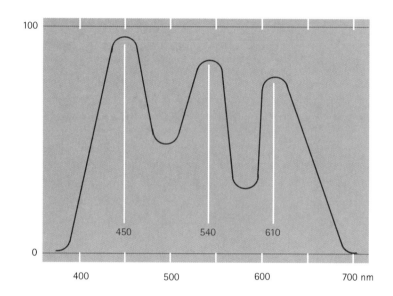

Wavelengths of visible light measured in nanometers with red (450nm) at the left and blue (610nm) at the right *(above)*.

Perception

When light reaches the human eye, two sets of nerve sensors within the eye react. Those called rods detect the intensity of light, and others, called cones, analyse the colour of perceived objects into a mixture of red, yellow and blue tints. This information is transmitted by the optic nerve to the brain, where the image of what we perceive is further modulated and recreated 'in the mind's eye'. The process of seeing involves vision, analysis and memory, and so is a complex and sometimes curious process in which we can be fooled by what we think we see, while what we actually see is overlaid by knowledge of what we are seeing.

The Hidden Giant. Popular print, 18C. A face in the landscape.

Perception and transformation: is this 18th-century popular print a landscape (horizontal) or a face (vertical)? *(above)*.

Colour

Colour is normally defined according to three categories: hue – the overall colour (red, blue or green, for example); value – the overall lightness or darkness of the colour; and chroma – the strength of the colour, its degree of purity or dilution. Thus a strong dark green would be described as having green as a hue and a high chroma and value, a pale washed-out red as having red as a hue, and a low chroma and value. Various systems exist for classifying colours along these lines, of which the Munsell system is probably the best known. This gives colours a three figure definition according to the values of hue, chroma and value.

Munsell's system is a splendid attempt to reduce the complexities of colour to a number of simple variables. The result however can only be an approximation, since colour is a function of the different wavelengths of light that are reflected by any particular object or surface, itself as we have seen, a factor dependent on the colour of the incident light. While we can also describe colours in terms of their wavelengths in the spectrum, these wavelengths are not fixed points; the wavelength is a ratio, and any change to the wavelength, however subtle, changes the colour of light or object perceived. Scientists have suggested that we can distinguish between up to 40 million colours within the visible spectrum, and what this actually means is that within a range of variations there are 40 million points at which the human eye can separate one from another.

The number of colours an individual can distinguish depends on the receptivity of the individual's own rods and cones, and is further conditioned by our subjective terminology for colours. You might describe a particular colour as an orangey-red, I might call the same colour a very reddish-orange. And there are even words whose meaning is now lost (Homer describes the colour of the sea as 'wine-dark' – a beautiful expression, but what did he mean?), while in some languages there are no terms for certain colours and shades at all. So one of the challenges of working with light is that an objective physical phenomenon, that can be measured with great accuracy, is in fact subject to a range of individual, social, cultural and psychological perceptions that may vary greatly from place to place and time to time.

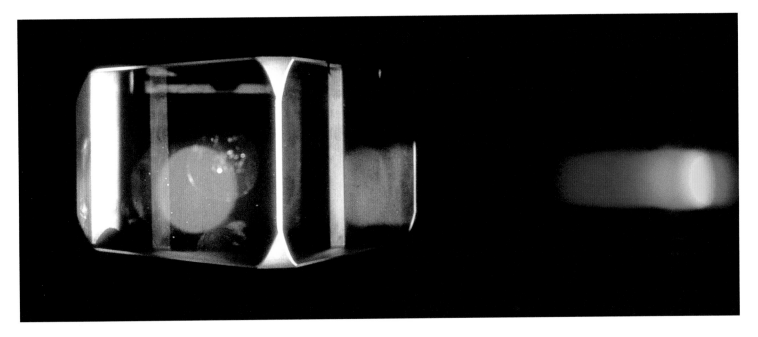

A beam of white light is split into colours by a prism *(above)*.

Colour of Light, Colour of Objects

Light is generally white, but passing a beam of light through a glass prism shows (and showed Newton) that white light is composed of the complete range of colours. (Very few light sources, natural or artificial, display the whole range of colours in equal proportions.) When white light strikes a coloured object, the surface of the object absorbs part of the spectrum and reflects the rest according to its colour. Thus a red object absorbs all wavelengths of light except red, which is why we see it as red in a white light. If the original light is not colour balanced, this will in turn affect the perceived colour. A red object seen in blue light appears black, for example, since there is no red light to reflect. The physical quality of the surface of the object also dictates the proportion of incident light it reflects or absorbs. This is termed the reflectance of the object or surface. Establishing correct colour values with lighting therefore has three component elements: the colour of light from the source, the innate colouring of the object or surface, and the reflectance of the object or surface.

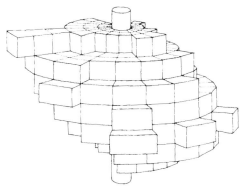

Munsell's colour shape is a three-dimensional colour model with a central axis representing chroma, and projections from it representing hue (according to position around the axis) and value (according to the length of the projection) *(above)*.

Sources of Light

For many centuries, and indeed still today, the most widely available source of light was sunlight, directly during the daytime and reflected from the moon at night. While sunlight can change in intensity and colour, because of atmospheric conditions such as the weather, and depending on the viewer's latitude on the globe, sunlight still remains firmly the standard by which we judge lighting effects. It has been argued that the brain applies a kind of sunlight constant to analyse scenes which would otherwise be underlit (for example at dusk) in much the same way that we can look at and read monochrome images such as films and photographs, without assuming that our friends or favourite Hollywood stars have suddenly all got stark white rather than skin-coloured faces, and that their wardrobe has become monochrome.

Artificial light sources began with oil lamps and candles both of which emit a redder, warmer light than sunlight, and it was only in the 19th century that more sophisticated light sources developed, initially with gas lighting, and then in the 1880s with the first electric lamps. (The principle of the incandescent electrical lamp was developed independently by Thomas Edison in the USA and Sir Joseph Wilson Swan in Britain in 1880.)

The candle's historic role as a supplier of light survives in the nomenclature of lighting technology: for a long time light output was measured in candle power. The modern metric term is luminous intensity, a measurement of the power of a source to emit light, though the term foot-candle is still used in the USA. The measurement of light follows four stages: measurement at the source, while light is flowing through space, upon its arrival at a surface, on its reflected return from the surface. The candela is the modern unit on which such calculations are based, calculated as one lumen of output for a solid conical angle or steradian at the source. The level of illuminance at a surface is calculated in lux, or lumens per square metre of the surface.

The same space by night and day: a tent created
for a temporary exhibition outside the Tate Gallery
in London *(above)*.

Colour and Light

Light sources do not emit an even white light, as anyone who has looked at different artificial light sources will know. The best known measure of the colour of light is colour temperature. This takes as a model an ideal surface which changes temperature when heated, and according to the light emitted the temperature is calculated. So a cloudless summer day would have a colour temperature of 10,000 degrees Kelvin, a dull afternoon about 4,000 degrees Kelvin and a cloudy sky, around 6,000 degrees Kelvin. Colour temperature is therefore a measure of intensity, and the formal analysis of light colour is achieved by analysing the spectral composition of the light from a source by plotting the wavelength of light on a graph. This relationship between colour rendition and temperature can be seen in the fact that low temperature sources tend to be weighted towards the lower, red end of the spectrum, while high temperature sources are found towards the higher blue end. The logic of this is confirmed by our experience; afternoon sunlight seems to have a red tinge, while bright light at noon has a blueish cast.

The lighting designer does not need to know the detailed mathematical formulae for calculating these values since they are available in technical manuals from suppliers and lighting associations, but a general understanding of the principles involved in measuring light is a valuable aid in developing a successful design.

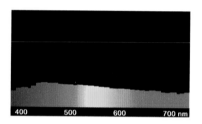

The spectral composition of natural daylight.

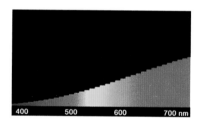

Typical spectral composition of an incandescent lamp showing its bias towards the red end of the spectrum.

Colour Rendition

When we say that an object is red, what is actually happening is the white light falling on it is absorbed except in the red part of the spectrum, which is reflected. The wavelength of light absorbed, and the intensity of energy reflected in fact depend, in the final analysis, on the chemical composition of the lit surface. We all know for example that a shiny surface, in whatever colour, reflects more light than a matt surface in the same colour.

If there is a change in the colour of the light source, then the reflected colours also change. A common example of this is the sodium lighting used on streets at night, in which the yellow light, while making shapes clearly discernible, flattens their colour range down, so that a blue object will appear black, or a white object yellow. Here our colour memory comes into play, allowing us to evaluate what we see as if it were better lit.

This phenomenon of colour rendition is extremely important for lighting designers. Different surfaces have their own colour values (the range of the colour spectrum they will absorb or reflect). They also have reflectance values (the amount of light – of whatever colour – they reflect). A wall painted in green gloss paint has a known colour value and high reflectance. The same wall covered in a green matt paint may have the same colour value with a much lower reflectance as the surface absorbs more light energy. The important variables for the designer are therefore light colour and reflectance.

Typical recommended reflectance figures for different building materials expressed as a percentage or as a decimal:

Glazed windows	0 (for practical purposes)
Granite	0–15 or 0.1–0.15
Dark stone	5–10 or 0.5–0.1
Yellow brick	35 or 0.35
Middle stone or medium concrete	40 or 0.4
Portland stone	60 or 0.6
White marble	60–65 or 0.6–0.65
White plaster	70 or 0.7
White brick	80 or 0.8

External weathering or urban grime can reduce these figures still further.

Artificial Light

Artificial light, from the candle flame to the low-voltage tungsten dichroic, the triphosphor fluorescent or the simple incandescent globe lamp used in many homes, is produced by using energy – normally electrical – to create light. On the following pages you will find an analysis of how contemporary lamps use electricity to create light, together with the basic classification of lamp types. The basic principle involves passing electricity through a space or a metal coil within an inert gas or vacuum, which translates that energy into light – and some of the energy into heat. To operate, a lamp needs to be placed in a fitting. This can be a simple lamp-holder or a complex device containing lenses and reflectors to guide and define the output. (Some contemporary lamps also have inbuilt reflectors and diffusers.) A 'light source' or light is therefore the combination of a lamp (or lamps) and a fitting (and any necessary control gear, such as a transformer).

In deciding on the appropriate light source to use for part or all of a lighting brief, the designer needs to have certain basic considerations in mind. Some of these relate to the practicality of the task, such as output, efficiency and cost, some to the aesthetic appearance of the lit space, such as light distribution, intensity and diffusion. Each of these will be looked at in turn after the basic lamp types have been identified. These different elements must be integrated into the final solution which will not depend entirely on lamp types, important as these are, but more often than not on devising a system that uses different lamps and fittings to achieve an overall result. In the case of museum and exhibition work, as we shall see, the ultraviolet component of light, and the heat emitted by lamps and fittings, are also very important considerations.

Discoveries in lighting techniques by scientists have helped artists – who repaid the compliment, as in Joseph Wright's *Experiment with an Air Pump (above)*.

Lamp Types

For practical purposes there are three main categories of lamps. These are fluorescent, including compact fluorescent, incandescent, and discharge. The classification depends on the electrical system used to create light, either by passing a current through a wire filament, or through an envelope filled with reactive gas.

A compact fluorescent low voltage lamp *(below)*.

A standard mains-voltage incandescent lamp with a silvered crown *(facing, left)*. A PAR 30 *(facing, centre)* and a PAR 20 *(facing, right)* tungsten mains voltage lamp.

Fluorescent Lamps

Fluorescent lamps operate by passing an electrical current through a gas or vapour contained in a glass bulb, which is in turn coated with phosphor on the inside. The electrical discharge created within the vapour (often mercury vapour) is transformed into light energy by passage through the phosphor coating. Such lamps are efficient in that a good proportion of the discharge energy is turned into light, and modern developments which use a triple coating of phosphor have allowed these lamps to give much better colour rendering. Fluorescent lamps also have a good lamp life.

The new range of compact fluorescent lamps offers the performance and life of conventional fluorescents, with the added advantage of a small lamp size. They are increasingly popular in many design applications, and a wide range of fittings is now available for them. Miniature fluorescents allow excellent lighting from a lamp little bigger than a pencil.

Standard fluorescent lamps require ballasts and starter gear; today electronic high-frequency ballasts should be used for preference since these avoid problems of flicker on start-up, whilst also allowing the lamps to be dimmed.

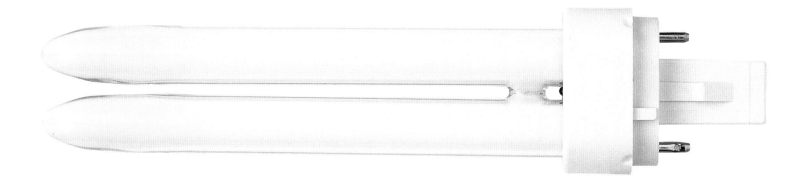

Incandescent Lamps

These lamps generate light by the passage of an electric current through a wire coil mounted in a vacuum. These were the earliest type of lamps to be developed and are still widely available, particularly for domestic use. They are, however, relatively inefficient, much of the energy used being radiated as heat rather than light, and the lamp life is relatively short. The first major development in incandescent lamps was the tungsten GLS lamp, which has a warm colour, followed by the PAR lamp, with its integral reflector, which allows better directional control. PAR lamps also have a longer lamp life. Tungsten halogen lamps have a light quality that is closer to daylight than standard or tungsten incandescents, as well as a longer lamp life. New developments in mains-voltage tungsten halogen allow these lamps to be retrofitted into existing mains systems.

Lower voltage tungsten halogen lamps offer excellent colour rendition, small lamp sizes, long life and low operating costs. However, these lamps require transformers, which are either integral to the fitting or need to be mounted near the lamp.

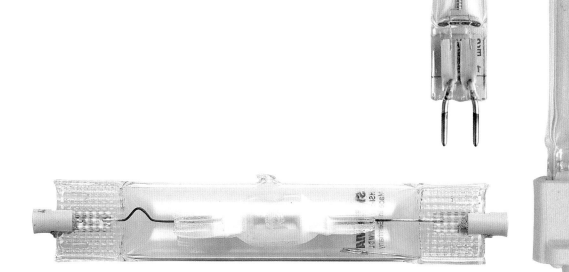

A double-ended mains voltage lamp *(above)*. A tungsten halogen axial filament low-voltage capsule lamp *(above, right)*. A metal halide lamp *(above, far right)*.

Discharge Lamps

These lamps are very efficient, whether in low- or high-pressure models, but require considerable control gear and cannot always be used in alternative burning positions. High intensity discharge lamps are based on mercury or sodium vapour as the discharge medium; sodium lamps tend to have an orange-white light colour and mercury a blueish colour.

Low-pressure sodium lamps emit a flat yellowish light with poor to no colour rendering: they are not appropriate for indoor use.

New lamps are always being developed: miniature fluorescents, mains-voltage halogen lamps, CDM-T lamps, ES 50 lamps and mercury-free high intensity lamps are also now available, the last an important development for ecological reasons.

Summary Guide to Lamp Types

	Type	Wattage	Life in hours	Colour temperature
Mains-voltage (240 volts) tungsten	Standard GLS	25–200 watt	1,000	2,700
	General purpose incandescent with good colour reproduction.			
	Reflector lamps	25–150 watt	1,000	2,700
	Mirror reflective coating on the inside of the lamp creates directional light in an uniform beam at angles between 25 (narrow) and 80 (broad) degrees.			
	PAR 38	60–120 watt	2,000	2,700
	Controlled light dispersal (12–30 degree beam angle) with high mechanical strength.			
Mains-voltage (240 volts) halogen	QT Capsule	75–300 watt	1,500–4,000	2,900
	Quartz halogen lamps for high output (luminous flux up to 5,000).			
	Halogen DLX	75–100 watt	2,000	2,900
	Much whiter light than standard incandescent with double the lamp life.			
	Halogen PAR 20	50 watt	2,000	2,900
	Halogen PAR 30	75–100 watt	2,500	2,900
	Halogen PAR 38	75–150 watt	2,500	2,900
	Can replace standard PAR or reflector lamps with higher output halogen.			
	QTY-DE	150–500 watt	2,000	2,800–2,950
	Luminous flux up to 9,500 from a double-ended tube.			
	ES 50	50 watt	2,500	2,900
	Small footprint but excellent output from integral reflector.			
Mains-voltage (240 volts) fluorescent	Tubular fluorescent	18–70 watt	7,000	2,900–4,300
	Colour temperature depends on lamp colour.			
	Triphosphor fluorescent	18–70 watt	7,000	2,700–6,000
	Triphosphor lamps are recommended for applications where accurate colour values are necessary.			
Mains-voltage (240 volts) compact fluorescent	Lynx s/se	5–11 watt	8,000	2,700–4,000
	Compacts offer the advantages of tubular fluorescent in a smaller format.			
	Lynx d/de, l/le, f	10–55 watt	8,000	2,700–4,000
	These lamps require starter gear.			
	Mini-lynx	7–20 watt	10,000	2,700
	Contain integral starters, so can be retrofitted to standard mains lamp-holders.			
Low-voltage (12 volts) halogen	QT	20–100 watt	2,000–3,000	3,000
	Compact dimensions with good output and colour rendering.			
	Dichroic	20–75 watt	2,000	3,000
	Well-defined beam (angles 8–60 degrees), excellent colour rendition, especially useful in food and heat sensitive contexts as 70 per cent of heat generated is radiated backwards.			
	Metal reflector	20–50 watt	2,000	3,000
	Aluminium reflector lamps with precisely directed beam (angles 6–32 degrees).			
High-intensity discharge mains-voltage (240 volts)	HIT/HIE metal halide	35–150 watt	5,000	3,000–4,000
	Good output with low colour rendering, require starter gear.			
	HIT PAR 38	100 watt	7,500	3,200
	Metal halide lamp with PAR integral reflector.			
	HST SON	50–150 watt	12,000	2,100
	Output equivalent to standard GLS lamp.			
	CDM	35–150 watt	6,000	3,000

Fitting Types

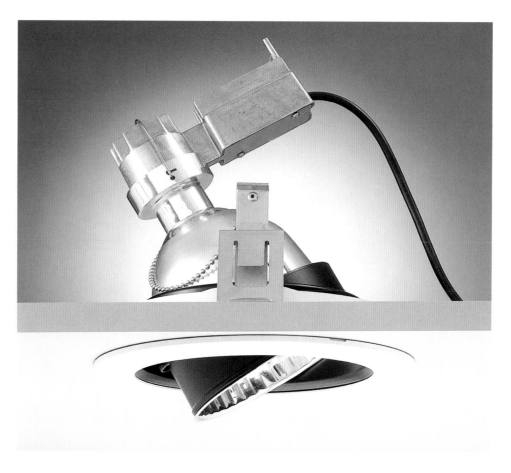

A light fitting performs a number of functions. It enables the electrical connection to the lamp itself, it protects the lamp, and it directs or diffuses the light from the lamp. In an indoor situation, the most important aspect of a fitting is the way in which it controls the flow of light. While fittings can be broadly divided into downlighters, uplighters, spotlights, ceiling-mounted, suspended and recessed fittings, many sophisticated modern fittings can be used in alternative positions, and combine qualities from other types, just as many lamps are configured in ways that in themselves direct the flow of light.

In many cases the fitting can be used with an alternative range of lamps. For example, a standard spotlight fitting will often accommodate lamps with different beam widths: here the work of controlling the flow of light is done by the lamp itself. Additional features available for fittings include diffusers, often used over fluorescent lamps, slots on recessed fittings, and plain or tinted glass rings to achieve an alternative diffusion of light from a recessed fitting.

In understanding the roles different fittings play in a complete lighting scheme, a useful starting point is the distinction between ambient lighting and accent lighting. Ambient lighting refers to the general level of lighting in a space, which may be from sunlight through windows, from a central pendant lamp, from striplighting in the ceiling, or whatever. Accent lighting is lighting that highlights details or particular areas. Often spotlights are used to provide accent lighting, since beam position, spread and intensity can be easily controlled and adjusted with spots. But wallwashers could equally well be used to highlight a wall-mounted display, for example, or downlighters set into a pattern that created stronger areas of light at floor or worktop level.

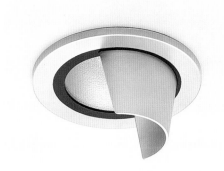

High-performance Equinox gimbal downlighter
(facing); Torus spotlights using 50W dichroic *(top left)*
and 100W capsule lamps *(top right)*; Myriad low-
voltage miniature downlighters *(above)* with fittings
including edge diffusers *(left)*, kick deflectors *(centre)*
and adjustable spotlight *(right)*.

Output, Efficiency and Cost

The output of a lamp is measured in lumens (in the UK and Europe) or candela (in the USA). This is a general measure calculating the light output, and typical values for different lamp types can be found in the chart on page 35. Where a very precise calculation of output is needed, the manufacturer's own documentation should be consulted.

Efficiency refers to the proportion of energy consumed by a lamp that is emitted as light, and also to the ability of the lamp to maintain consistent output. Modern lamps are generally very efficient in this latter respect, maintaining an even output over a longer percentage of their lamp life than was the case earlier.

Cost can be calculated in two ways. Firstly, there is the capital cost of a lighting installation, representing the price of fittings, control gear, transformers, lamps, etc. Then there is the running cost of the installation, represented by the electrical consumption, and the cost of relamping. There are various systems of cost comparison, but a rule of thumb is to divide the capital cost by the likely lifespan of the installation and add this to the running cost in order to establish a yearly total.

Relamping can be either carried out on a formal basis – all the lamps in an installation being changed after a fixed period – or can be done *ad hoc*, that is to say lamps being replaced as and when they fail. Both systems have their advantages; the main consideration for the designer is to ensure that access to lamps that may need frequent changing is not too difficult.

The colour of different lamps can be seen from these three images. Each was taken using the same intensity of light, but from different sources. *(Above, from left to right)* tungsten halogen, metal halide, compact fluorescent.

Light Distribution

Light distribution is a function both of the lamp and the fitting, since in most cases the fitting controls the direction and the power of the light source. The designer needs to understand the distribution pattern of a particular fitting and this is best achieved by looking at a diagram called a polar curve for the fitting in question. This plots the outline and intensity of light from a specific fitting.

Lighting angles affect distribution: the same object lit from above, behind and from the side appears very different, and the degree of detail visible changes *(left)*.

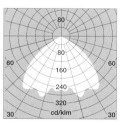

A polar curve diagram represents graphically the luminous output of a given fitting and lamp in different directions. The central column of numerals indicates the intensity in candelas per lumen, the outer columns, the angle of output in degrees. For most fittings a single plane is shown, but for a wallwasher or other highly-directional fittings, two planes, perpendicular to each other are represented *(above)*.

Intensity and Diffusion

The intensity of light falling on a particular surface, for example a worktop, sales point, or display is translated by looking at the incident light falling on it in lumens. There are guidelines available for the desired level of lighting in different work and retail situations; in some cases these are linked to official regulations so the designer needs to be aware of these.

Some recommended levels of lighting are as follows: the recommended level of illuminance in lux for small shops is 500 lux, for supermarkets and hypermarkets it is the same amount calculated on the vertical faces of displays, and for checkout areas also 500 lux calculated on the horizontal plane of the checkout or sales area. These can be compared with levels of 150 lux, deemed appropriate for areas such as loading bays, 750 lux for drawing offices, and 1,500 lux for hand-tailoring, operating theatres and suchlike.

The diffusion of light refers to the overall level of lighting achieved in a scheme by the complete installation. Computer programs are now widely available for calculating these results, and for testing alternative solutions for a specific interior space.

A typical interior program allows lighting levels to be calculated in a rectangular or L-shaped interior space *(below, left)*. The user selects and positions the luminaires in the space from the available database and the program produces a visible output of the lighting distribution *(below, right)*.

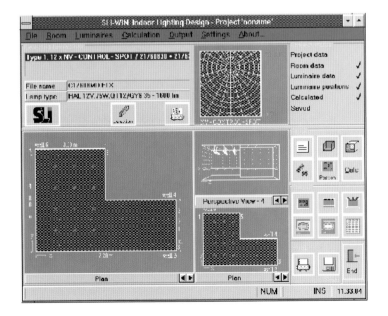

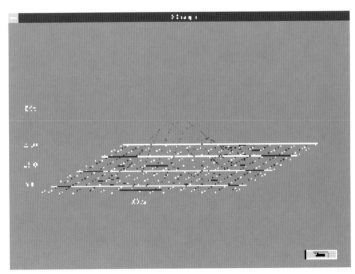

Two images of virtual museum spaces, but with correct lighting effects, generated in Lightscape *(above)*.

To calculate lighting values exactly, lamp and lighting manufacturers have created programs that will produce an outline of lighting levels in a regular space from exact technical data. These allow the architect or designer to check that lighting distribution and levels meet, for example, health and safety requirements. However, such programs do not produce a rendered image, so the visualisation process is incomplete. High-end graphics programs can create visually appealing images, either from CAD data or from the designer's eye, and technical programs can describe the lighting solutions exactly. What was needed was to marry the two up, so that the designer could show a client a correct image of the project.

Two recent programs have squared this particular circle. The first, Lightscape, is a stand-alone program. It offers a series of libraries of surfaces, elements and light fittings that can be assembled to replicate an interior and then lit. The complete radiosity program calculates the lighting effect accurately and from all positions. Lightspace recently released a link for 3DStudioMax and the

new 3DStudioViz that allows the program to import files from 3DStudio and render them using the Lightscape radiosity engine. Lightscape are also regularly releasing further libraries that contain both architectural elements and specific light fittings to extend and enhance the range of luminaires that can be used under Lightscape.

RadioRay, developed for Autodesk by the UK company LightWorks, is a plug-in for 3DStudioMax and the new 3DStudioViz. This also allows precise light fittings to be inserted into interiors created in 3DStudio, using the industry standard .elx file extension. These files, created by lamp and fitting manufacturers, contain complete photogrammetric data allowing the exact performance of a lamp and fitting to be depicted. So in a finalised and rendered design, precise lux values can be read off walls and work surfaces, and the colour and diffusion of light are also visually accurate. Again, since this is a radiosity program, the values are created for all viewpoints.

Light and Visual Effect

Thus light colour, angle of incidence, and spread of light on an object will have a radical effect on how the object is seen. Something lit directly from overhead looks very different to the same thing lit from below or at an angle. And when similarly the same object is seen lit by different light sources it will appear with very different colour balances and even rendering of detail.

Lighting angle and beam width changes the appearance of this sculpture radically *(right)*.

Lighting and Conservation

The ultraviolet component in sunlight, as mentioned earlier, has an important consequence for historic objects – particularly paintings, works on paper, costumes and natural materials – and where these should be exhibited or displayed. Too much exposure to light can cause damage and fading to light-sensitive objects. Museum curators and lighting designers have therefore divided exhibits into three categories. The first category, unfortunately the smallest one, consists of those objects whose component materials are insensitive to light; examples of this would be unpainted stone, metals, ceramics and jewellery. In the second category are objects which are moderately sensitive to light, often because the pigments with which they are decorated or painted change colour slowly under light. Oil and tempera paintings are the largest group in this category. A typical and serious case of light-induced damage is with a widely-used green pigment, which changes under exposure to light from grass green to look an opaque dark brown. The red lake pigments fade equally under light, and so flesh colours will often suffer in oil paintings if they are overlit. A satirical poem written at the time of the death of Reynolds, the founder of the Royal Academy in London, is a reminder that this is a problem which has been with us for a long time. The third category covers very light-sensitive materials. These include textiles, watercolours, works on paper and natural history specimens. Here light does not only cause colour changes but also weakens the material structure of the object. In effect, all things decay.

The problem facing the museum conservator is therefore one of the rate of exposure; ideally, if all a museum's watercolours, for example, were kept in total darkness, they would remain bright and fresh for ever, but would be of no benefit to anyone because nobody would ever be able to see them. The general approach is now based on three principles; firstly, we should seek to reduce illuminance to no more than is necessary for comfortable viewing. Secondly, the time of illumination should be kept to a minimum, as far as possible. And, thirdly, ultraviolet radiation, which is by far the most damaging component of light, should be reduced too. It is good practice to consider this first principle in terms of the lux/hours of exposure of an object (since 50 lux for 1000 hours is as damaging as 500 lux for 100 hours), and decide what level of exposure per year is acceptable for which categories of object. As a guide to this, category-two objects are normally considered to require a lux level of not above 200 and objects in category three a lux level of not above 50. In order to achieve this lower level successfully, the lighting designer must pay considerable attention to the progress of the visitor through the exhibition area. 50 lux, for example, is a relatively low level of light and thus it is important to provide space in which the visitor's eyes can become accustomed to the newer light level before the visitor moves into the new space. Lamps of low colour temperature have been found to be more acceptable to the eye at low light levels.

A Somerset Church by Mildred Drury. By courtesy of Gillian Hooker.

Evidence of damage by exposure to light can often be revealed when a watercolour or wash drawing is removed from the original mount: the parts under the mount have retained full original colour, the rest appears washed out *(right)*.

Some of the problems of integrating modern lighting into old buildings can be seen in this interior at the Louvre Museum (which has since been relit). Limited placement opportunities and large fittings did not help *(facing)*.

As to reducing exposure time, the simple and traditional device of curtains over showcases is one option. Another is a time switch operated by visitors, or an advance sensor that detects visitor movements and turns on lights before and after the visitor passes. It is also important to reduce light levels, either completely or to the minimum necessary for security, when galleries of exhibition areas are closed. Experience suggests that automatic systems are better than manual systems, since warding and other staff have more important tasks than controlling lighting, and may not fully understand the complexities of a complete system.

Reducing ultraviolet light from fittings is relatively straightforward. Either lamps with low UV emission should be chosen, or UV filters provided for existing lamps (such as linear fluorescents) with a high UV output. This can be either a plastic filter in acrylic or polycarbonate, or a flexible film or varnish layered on to existing glass. For quartz halogen lamps a plate glass screen is recommended to absorb the small quantities of UV they produce. A UV monitor to meter UV output is a useful addition to the lighting designer's kit, together with a standard light meter necessary for checking levels of ambient or accent light.

Ultraviolet light is also present in daylight, and this raises the great daylight dilemma. Many historical works of art were painted in daylight and intended to be viewed in daylight, or may hang in traditional settings where daylight is an important component of the general lighting. There are those who believe, with reason, that such works are best seen under daylight conditions, since that is being faithful to the original artist's intentions. And in an historic interior it may not be acceptable to damage the structure by inserting modern wiring and light fittings. If the designer has a choice in a daylit interior, the best solution may be to fit shutters or blinds to windows to reduce the level of direct light, especially from the upper sky, with a control system (for example using photoelectric cells and motorised blinds) to adjust the interior light to a safe level as the exterior light fluctuates through the day. That approach may also need some artificial lighting to supplement daylight, for example in winter, and to replace it in the evenings and at night. If excluding daylight completely is an option, it should be followed. Artificial light can be selected and placed to give a daylight effect, and the capital and maintenance costs of a wholly artificial system may well be less than for a complex controlled/mixed system.

A final consideration is the heat emitted by lamps and fittings. This can do as much damage to sensitive materials (and no good to insensitive ones) as light can. So it is important to check in placing fittings that they are not too close to exhibits, or where this is unavoidable to ensure a proper air flow to remove the heat, and to select lamps and fittings with the best heat characteristics. For example, within showcases fibre-optic light can be a useful solution since the light source can be outside the case, so eliminating the heat problem. New dichroic reflector lamps are available which direct the maximum heat output to the rear of the lamp rather than to the front. Provided the fitting is adapted to let the heat escape, this can be a useful solution.

Carefully planned lighting on an antique glass vase allows its quality and details to become clear, without any risk to the fragile material *(above)*.

For the lighting solution, see the diagram *(right)*.

The need to meet conservation requirements is an additional challenge for the lighting designer working with all works of art. Once this challenge has been met, the designer's task is to find a solution that allows the visitor to enjoy the work on display, achieving an ambience in which the work can be enjoyed by the visitor directly and to the full.

An ingenious system to reduce both incident light and heat from fittings used in a museum in Switzerland: the fittings are mounted on top of the display panels, and project on to movable screens that reduce both intensity and glare *(above)*.

Lighting as a System

It is important to think of all lighting solutions in system terms, as an approach to lighting that only looks at independent elements and not towards the total result is likely to be unsuccessful. A systematic approach does not mean using only system lighting: it means bringing together all the requirements of the design and the individual decisions on lamps and fittings into a coherent whole. A successful solution will meet the immediate lighting needs of the space, be simple to operate (in terms both of relamping and switching different parts of the system on and off) and be able to be adapted to meet new display requirements for new works. (Even in a permanent collection, the work on display will change). A track system, whether ceiling-mounted or suspended, is the most obvious example of how to create these opportunities for change, and track systems are widely used for this reason. But fixed-position lighting can also have inbuilt variability, through the careful planning of circuits to allow for different combinations of sources to be lit together or in groups, and through choosing fittings that can accept different lamps, with alternative outputs, beam widths or light colours, and which can be modified by additional elements such as diffuser rings, barn-doors or filters. The designer's job is to make the visitor's experience as rewarding as possible, and this is often best attained by allowing for the future while meeting the demands of the present.

The redesign of the public areas at the Royal National Theatre in London by architects Stanton Williams uses a specially designed four-lamp downlighter from Concord, together with specially adapted chromed theatre lamps by Marcus Brill *(facing)*.

There are two main practical challenges facing the lighting designer seeking to light an exterior space. The first is the range of fittings available, the second, the limited choice of lighting positions. Fittings for exterior use have, evidently, to be weatherproof, resisting wind and cold, dew and rain. Cables and switches also need to be shielded. If the fittings are to be mounted within people's reach, they also need to be protected against both deliberate and accidental damage, and the consequent risk of injury.

This means exterior fittings are in general bulkier than equivalent light sources for indoor use. There is also the question of lighting positions for exteriors. Lighting can be mounted at ground level, either inset into the ground or mounted on stands or poles, creating uplighting or horizontal lighting. Or fixtures can be attached to the roof-line or edges of the building itself, to create a wallwashing

downlight. Both these solutions have their drawbacks and difficulties. A row of lighting poles may produce the right effect at night, but appear unattractive in the daytime. Traditional architectural details such as carvings and sculptures will often have been designed to be seen lit from overhead, not from below. And the condition and conservation requirements of a historic building fabric may preclude attaching fixtures to it.

The exterior of Terry Farrell's Charing Cross
building in London *(above)*.

The St Katherine's Dock area in London at night,
where angled wallwashing from ground level
reduces light pollution *(facing, above)*.

Lighting enhances the impression of a temporary
structure within the walls of the ancient fort at
Bellinzona *(facing, below)*.

There is, however, increasing interest in lighting modern and historic buildings, and open spaces. This comes in part from a sense of civic pride, from the need to make a corporate statement, and from a wish to improve the urban environment, in making the way safer for pedestrians and in deterring crime and vandalism. In this latter case, where the security aspect is a key part of the lighting brief, the lighting expert will need to talk to further experts so as to calculate the most efficient emplacements, and to co-ordinate lighting with other security equipment such as closed-circuit television cameras.

Even if the purpose of a lighting scheme is mainly functional, design values need not be abandoned. The lighting of an exterior should in general be through an even wash of light, just as the lighting of a pathway or road should be even – especially if there are steps of changes of level involved. However, coloured lighting (for example, for a corporate headquarters) can also be appropriate, and even lighting schemes that change the placing and intensity of light are possible.

The risk of accidental or even deliberate damage (for example from water penetration or vandalism) is a serious one for designers, as damage will not only render the fitting useless, but may also make it a danger to people using the space. Modern fittings are normally rated according to internationally-agreed standards of protection, showing how resistant the fitting is to potential risks. The IP numbers used by manufacturers to describe their fittings refer to these standards – an IP23 fitting is rainproof, for example, and an IP65 fitting proof against heavy blows. The designer should check the rating of proposed fittings, and be aware of any relevant fire or building regulations in making the choice of fitting. A further technical consideration is the accessibility of the particular fitting, once in place, for maintenance, cleaning and relamping.

Another important consideration is light pollution: light sources should light definite areas, and not simply spill light into the sky. This is both a waste of energy, a distraction for both aircraft pilots and astronomers, and proof of an ill-conceived design. Even though the planning and calculation of exterior projects is somewhat more difficult than interior ones, for example with the smaller range of appropriate fittings and the absence of reflecting surfaces, a successful exterior project can not only be elegant in itself but can contribute to the pleasure of a nocturnal landscape or streetscape.

Nocturnal view of the Issac Newton Institute in Cambridge, England *(above)*.

Frank Gehry's new Guggenheim Museum in Bilbao uses the reflective surfaces of titanium to maximum effect *(right)*.

Louvre Museum,
Paris, France.

'Recreating the subtlety of natural daylight was essential.'

I.M. Pei's glass pyramid in the courtyard of the Louvre in Paris is widely – and rightly – regarded as one of the great high-tech creations of modern architecture, particularly given its historic context. The pyramid was the first phase: the redevelopment of the Louvre was continued in 1995 with the creation of I.M. Pei's Richelieu wing and the opening of the Cour Carrée to the public, and in 1997 by the reopening of the newly laid out gardens between the Louvre and the Place de la Concorde. This created an alternative, pedestrian route between the river Seine and the rue de Rivoli.

As to the Cour Carrée, the French electricity board, EDF, proposed lighting this courtyard, originally designed by Le Vau in the late 1660s. The three-storey columnated façades are richly decorated with sculptures, and form one of the most significant historic buildings in Paris, as well as part of the site of one of the most important museums in the world. This was a context that required sensitivity, not only in handling the fabric of the building, but in presenting the complex elements (columns, fully-rounded statues and bas-reliefs) evenly. A further problem was that the sculptures were created to be seen under daylight, that is with an overhead light. It was not possible to build a projecting overhead lighting system, but too much lighting from below would create grotesque patterns of shadows.

Statue of Aphrodite in the Pavillon des Arts in the Cour Carrée *(facing)*.

The path to a solution was found by devising special computer technology in association with the Centre Nationale de Récherche Scientifique (CNRS) and the Mensi Corporation, which allowed a virtual three-dimensional model of the courtyard to be created from laser scans of the actual façades. This model was accurate both for colour and surface reflectance. Virtual light sources could be imported into this model and positioned at will, which then gave a rendered impression of the lit effect. This Phostére program allowed different lighting solutions to be evaluated in detail and in general. It is a measure of the importance and sensitivity of the project that this program was developed specifically for the task. The computer analysis suggested that a large number of relatively low intensity sources would provide the required

Photograph© La Médiathèque EDF/Claude Pauquet

effect better than a smaller number of more powerful lamps. Most of the lamps could be placed along the upper balustrades, but at a sufficient angle to light sculptures and bas-reliefs generally, providing a consistent and convincing lighting effect from ground level.

The Pavillon de L'Horloge with caryatids by Guérine de Buyster *(above)*.

Abundance or Nature by Jean Goujon on the Lescot Wing of the Cour Carrée *(facing)*.

The Law by Jean Guillaume Moitte, on the Lemercier wing *(above)*.

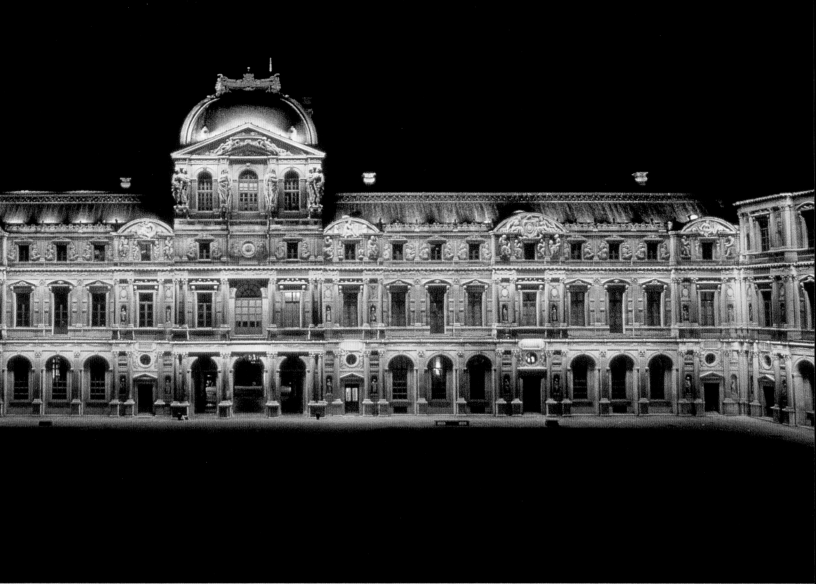

Incandescent light was chosen as closest to daylight, and a low-voltage xenon lamp in a specially designed reflector was used for the main installation, supplemented on specific sculptures by dichroic spotlights. Conventional spotlights were used to light the roof masses.

The result is not a simulation of daylight at night (what the French call, in homage to the film industry, *La nuit américaine*), but rather an independent and wholly original lighting pattern, reminiscent of the warm colours of torch and lantern light, but with no element of pastiche. It is a solution that not merely respects the historic fabric of the building with care, but also creates a wholly new impression of it as a lit space.

The Pavillon de l'Horloge in the centre of the Cour Carrée *(above)*.

Sculpture at Goodwood Visitor's Centre and Gallery, West Sussex, UK.

'Like Laugier's primitive hut, this is the outcome of similarly sophisticated rustic yearnings.'

A collection of contemporary sculpture to be displayed on a wooded hillside in the south of England: a balance of natural landscape and human artefact, and a space to enjoy and explore. These were the ideas of the private collectors and benefactors who created Hat Hill. They invited the architect Craig Downie to obtain planning permission for the 20 acres and build the Visitor's Centre, a small central building to act as a central point in the park, where visitor's could orient themselves among the work on display and also find out more about the artists, through displays of studies and models on an interactive CD-ROM library.

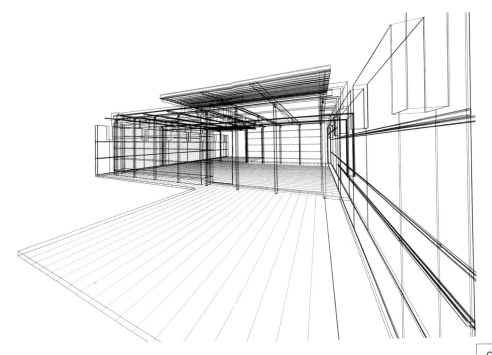

The Visitor's Centre in it's woodland setting *(facing)*.

Perspective computer image of the design *(right)*.

65

Exterior of the Centre showing uplighting and downlighting wallwashers *(above)*.

Interior view of the Visitor's Centre *(facing)*.

FIRE EXIT

While the building has a central weatherproof core that is the information and exhibition area, it was also designed to open out into the surrounding park, and act as a focal point for the pathways throughout the park.

The interior has a lighting system to supplement daylight on cloudy or dull days, and a deliberate overspill to mark the position of the building at dusk. A track system within the wooden ceiling allows flexibility for changing displays, with fixed-point floodlights to illuminate doorways and access points.

View from the outside showing external wallwashers and internal tracked spotlights *(above, left)*.

The interactive display area in the Centre *(facing)*.

Computer generated elevation of the Visitor's Centre *(below)*.

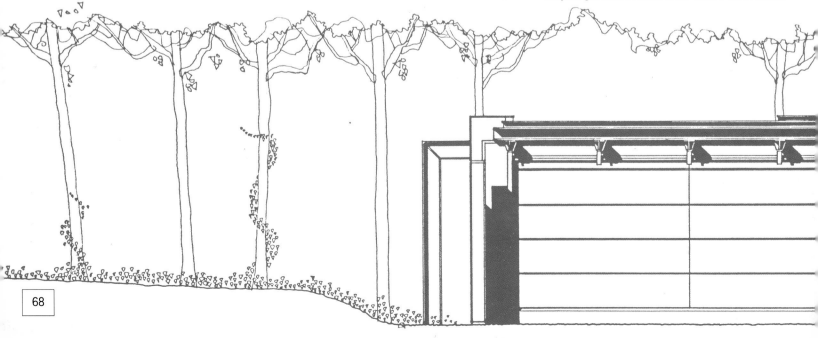

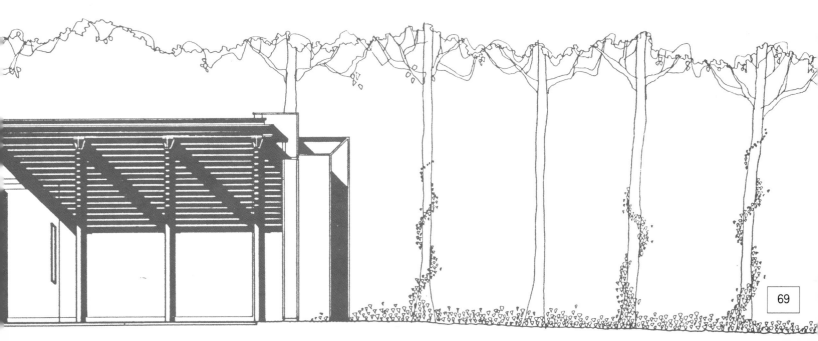

'Gas light' wrote the novelist Edgar Allan Poe in 1895, 'is totally inappropriate for the interior. Its harshness and inconstancy is painful. No one with vision and common sense will use it.' Other 19th-century critics were more severe: Ludwig Borne asserted that 'gas light is too pure for the human eye, and it will cause blindness in our grandchildren.' With one hundred years of experience of artificial light, we now know that, correctly used, it is neither harmful nor inappropriate.

We have the technology to moderate its 'harshness and inconstancy'. The task for the lighting designer, in illuminating a historic interior or lighting an exhibition of works of art, is to use this technology wisely: to add the understanding to the technology. In an interior, one important question, once the conservation issues have been addressed, will be to imagine what the original lighting levels of the interior were, and to have consideration for these in any new scheme (but not necessarily to imitate them inflexibly). And in an exhibition, to follow the lead given by the curator or organiser in selecting appropriate light levels overall, and for which objects or works to highlight.

Public art galleries and exhibition spaces are, architecturally speaking, a relatively new phenomenon. The English landed gentry or the European aristocracy may, from time to time, have built room to house and display works of art, such as the sculpture court at Petworth or the old halls of the Louvre, but purpose-built public buildings for the visual arts are a fairly recent concept.

They only began to emerge in the mid-19th century, with many of the most exciting examples dating from the last 50 years of the 20th century, in creations such as the Guggenheim Museum in Bilbao, the Pompidou Centre in Paris, and the Menil Gallery in Houston, Texas. Also in recent years, older buildings have found a new lease of life through use as museums and exhibition galleries, from the lofts of New York's SoHo to a multimedia centre in a fortress on the Seine or a Spanish factory housing a collection of glass. With new categories of objects from printed ephemera to primitive art coming into the museum area, and through new interest in old pursuits and contemporary awareness of popular culture, the diversity of exhibited work is changing and expanding all the time.

The 'Genius of Venice' exhibition at the Royal Academy in London, designed by Alan Irvine *(facing)*.

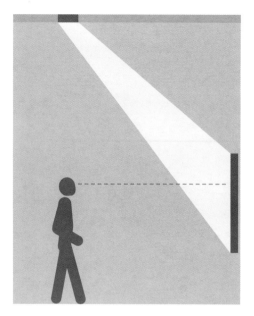

Correct lighting of pictures depends on lighting them at the correct angle for the viewer *(left)*.

The aim of any exhibition, temporary or permanent, is to delight and inform the public, be they expert or amateur. A well planned, well hung and well lit exhibition can be a visual joy, whether the work displayed be new or familiar. The lighting designer, with the other members of the team, builds the visual bridge between the visitor and the work that is essential for a successful exhibition. In analysing how this is achieved we will begin with surface works such as paintings, drawings and engravings. In terms of the lighting of an individual work, the first aspect to consider is the risk of glare. This is particularly the case when a work of art is framed or mounted under glass, where the glass surface reflects incident light into the viewer's eyes, masking the painting. The second factor is the even fall of light across the whole picture surface, so that there is no variation in the illumination of each area. A general starting point for a solution is to calculate a source position for the light by imagining a line running from the centre of the painting (or work) to the ceiling at an angle of 35 to 45 degrees from the vertical plane of the work, and placing the fitting on that line, whether at the ceiling or below it, with its beam directed at an angle of 55 to 45 degrees to the ceiling plane. Detailed factors affecting the angle will include the depth of the frame or surround of the work (a deep frame can cast shadows on the picture surface if the light angle is too acute), and the size and hanging position of the work itself.

Artists talk about being 'hung on the line' at important exhibitions. Ideally, a painting should be hung so that a horizontal line one third down the painting is on the level of the viewer's eyes. Some galleries even painted a line at this height to act as a hanging mark. The best paintings would be hung to this line. (Paintings hung far above the line in a crowded show were said to be 'skied'.)

In this display from a temporary exhibition, the designer has balanced the light on both the paintings and on the descriptive texts. Given the different reflectances of the surfaces, this requires a high degree of skill to work properly *(above)*.

Getting an even wash of light will depend on the choice of fittings and the spacing between them. The same applies to lighting a series of items hung on the same wall: once a general lighting position has been found the designer must decide whether to light all the works as a single group, or to light them individually. If the works are lit individually, it is important that the change of light level between each picture in a row is not too abrupt so that the viewer can make a visual comparison, and of course that the light intensity and colour is the same on each work.

This group of silk-screened images by Andy Warhol, shown at the Tate Gallery in London, have been treated as a single item though lit from a row of sources. This balances the treatment of the large scale portraits on the next wall back *(facing)* and so guides the viewer from exhibit to exhibit *(above)*.

There is an alternative approach which lights the individual object independently of its context. This often leads to a solution where all the light is on the exhibits with relative darkness around. This can often be a useful approach from a conservation aspect when light levels need to be kept low, and can be appropriate where works need to be hung in a group but not necessarily seen as a group. (This might apply in a mixed historical exhibition with a group of paintings by different artists all painted in the same year, for example, or for the same patron. The works would make individual statements and an overall one, but not necessarily comparative ones.) Among exhibition designers there are some who feel that a uniform, mid-toned or even dark background colour is the appropriate setting for works of art, and in such a case the lighting designer will have to work out carefully the impact of the chosen colour on the ambient light levels and in turn, on the accent lighting on pictures displayed.

A group of Russian icons lit independently against a dark wall colour in an exhibition at the National Gallery in Manila *(above)*. 100-watt parabolic spotlights have been used, but the incident light on the paint surfaces is only 50 lux. Direct rather than general lighting has helped to achieve this.

The National Gallery in Berlin, designed by Mies van der Rohe, uses a range of wall colours and a general approach of reducing light in the main area to concentrate light on the pictures *(above)*.

There was no major conservation issue here, rather the aesthetic concept of a secular temple to 20th-century art, with only downlighters and wallwashers, both ceiling-mounted, providing the lighting. Note how the solution works through to the exterior at night *(left)*.

The Atlantis Gallery,

London, UK.

'Giving contemporary art a sense of place.'

Contemporary art is big: not just in terms of the controversy it sometimes creates, but in terms of the scale of work many contemporary artists produce. The range of materials has moved from paint and canvas, stone and bronze, to add industrial articles, mechanical and human processes, even animals. Creating an exhibition space for a range of contemporary works would therefore appear to be simple: a large neutral box would be able to handle anything. In fact, it is far more complex than this. A space, especially one for changing exhibitions, has to have in a sense a personality of its own, something which the regular visitor can relate to, a constant amid the changing works. This may be no more than an architectural feature such as a doorway or roof-line, or a sense of proportion and space. And the quality of light can contribute to this 'sense of place'.

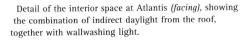

Detail of the interior space at Atlantis *(facing)*, showing the combination of indirect daylight from the roof, together with wallwashing light.

The Atlantis Gallery in east London was converted from the enormous storage loft of a nearby brewery (breweries seem to make good galleries – the Museum of Modern Art in Oxford is also situated in a former brewery building). Storage lofts have certain advantages: solid flooring with no weight problems, and uninterrupted central space, often with, as here, only occasional pillars, and a good ceiling height. At Atlantis the concrete floor has been polished, the walls and pillars painted white, the roof trusses and frames painted in black. The floor area is a simple rectangle of 14,000 square feet, and the double-pitched roof at its highest point is 23 feet above the floor. This means the roof can add character without detracting visually from even the largest works on display on the floor or walls. The ambient lighting comes in during the day from four banks of roof skylights supplemented by concealed fluorescents. Accent lighting comes from track-mounted low-voltage spotlights. The track is midway between the walls and central line of pillars, giving the correct general angle for lighting paintings or other work on the walls.

Detail of the interior space at Atlantis *(facing)*, showing wallwashing light over exhibited work. The fact that contemporary work is mainly shown unframed allows a close angle for wallwashing. With traditional, framed paintings, care must be taken to avoid shadows from the frame falling on the picture edge, especially when using wallwashing light.

Duveen Wing, National Portrait Gallery, London, UK.

> '*A series of fluid and uncluttered spaces in which portraits and sculptures can be fully appreciated together.*'

Joseph Duveen, who died in 1939, was one of the great art entrepreneurs of the 20th century. Realising, as his biographer Sidney Behrman said, that 'Europe had plenty of art and America plenty of money', he set about enabling wealthy Americans to become collectors and patrons, making a fortune in the process, much of which he bequeathed or gifted to museums in Britain, including the National Portrait Gallery in London, of which he was a trustee. The wing in which 20th-century portraits are shown is named after him.

In 1994 it was decided to redesign the Duveen wing, and relight the adjacent Victorian wing used for 19th-century portraiture. Piers Gough of the architectural practice CZWG was appointed to head the team, which included Ove Arup as engineers and Concord Lighting as lighting designers. An important part of the brief was that portrait paintings and sculptures were to be exhibited together. The galleries reopened in 1996 to considerable acclaim.

'We wanted to provide the galleries with at least two lighting positions for each wall surface, between the ideal angles of 20 and 45 degrees to the picture plane,' Gough explains. 'We also wanted to avoid the linear emphasis that continuous track lighting tends to produce, and the clutter of fittings other than spotlights that are required in a modern gallery.'

General view of the Duveen wing. Ambient lighting comes from daylight, with accent lighting framed on to the pictures from low-voltage halogen spotlights *(facing)*.

The existing rooms of the Victorian Gallery were reorganised (particularly by moving the location of the lecture theatre) to provide more hanging space, to ritualise windows and to create a clearer pattern for visitor movement. The room ceilings contained a fine grid of plaster mouldings, and the distance of this grid from the walls was found to be compatible with the preferred lighting positions. Gough therefore conceived the idea of a lighting grid dropped below the ceiling plane, as a reflection of the mouldings and to protect the ceiling, and avoid a forest of suspension wires and supply cables. The track is supported by rods at each corner of the grid: 'a difficult detail to achieve on site' as Gough points out.

In the Duveen Wing the architects had a much freer hand. It had previously had a low ceiling, blocked up windows, and angled partition walls. Under the redesign, all this has changed. The windows have been reopened, the ceiling lifted and curved, the old partitions replaced by glass walls, a radical alteration that provides a visually exciting, and naturally lit, high-ceilinged single space. To achieve this there were a number of challenges for the design of the lighting. 'Firstly,' Gough explains, 'in creating a new ceiling we wanted a visually clean soffit without conventional track or visible fittings. Secondly, we wanted to have two lighting positions for each glass wall, at the correct angles. And thirdly, we had to avoid light from the fittings passing through the glass walls and distracting people looking at the pictures.'

The solution found was to make a series of slots in the ceiling, running parallel to the glass walls, perpendicular to the window wall. These slots enable all the fittings and track, as well as CCTV, smoke detectors, emergency lighting, alarms and P.A. speakers, to be concealed from view. In turn, the slots are positioned at a frequency that permits two lighting positions for each glass wall based on the 20 to 45 degree lighting angles mentioned earlier. And thirdly, to avoid light spillage and glare, the light beam had to be controllable: spotlights with framing heads, mounted on short lengths of track were the answer. The tracks allowed different source positions and left or right orientation on to the adjoining glass walls, and the framing heads enabled the light beam to be adjusted so as to fall only on the picture and frame, without any overspill on to the surrounding glass surface. Finally, an automatic system of blinds has been fitted to the windows to control daylight.

This spectacular redesign has given the new Victorian and 20th-Century Galleries a series of fluid and uncluttered spaces in which portraits and sculptures can be fully appreciated together. A subtle system of holes and brackets in the glass walls allows for changes in the display, which in turn can be accommodated by the lighting. The sinuous curve of the fibrous plaster ceiling creates a visual impact without distracting from the work on display.

The Victorian Gallery as relit. By adapting the normal track fittings to a horizontal system, the track can be visually integrated into the ceiling without suspension wires. Track also allows for changed lighting positions. These spots use 20-watt medium-beam dichroic low-voltage lamps *(previous page)*.

Two views of the Duveen room before hanging began. Note the glass planes with multiple fixing points for attaching pictures *(left)* and the lighting system recessed into ceiling slots *(above)*.

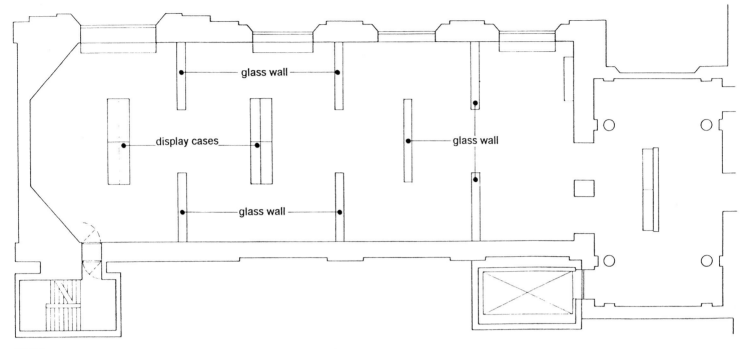

PLAN

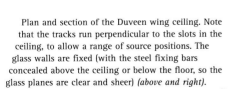

Plan and section of the Duveen wing ceiling. Note that the tracks run perpendicular to the slots in the ceiling, to allow a range of source positions. The glass walls are fixed (with the steel fixing bars concealed above the ceiling or below the floor, so the glass planes are clear and sheer) *(above and right)*.

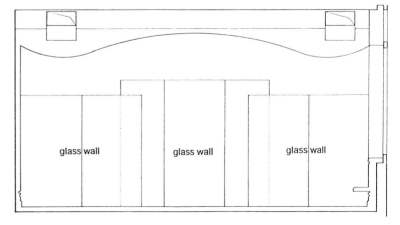

CROSS SECTION

Mondrian Exhibition, Tate Gallery, London, UK.

'Nature to abstraction.'

The Dutch artist Piet Mondrian is one of the most influential abstract painters of the 20th century. His formal grids of black lines and colour patches enunciate an inarticulate language of form, and have had an immense influence on other artists. But his earliest work was representational, and this exhibition, shown at the Tate Gallery in London in 1997, allowed the visitor to trace the artist's development, from his earliest landscapes of ditches and dikes, to his final, best-known grids. Displayed in three smallish rooms, most of the works in the show were on loan from the Haags Gemeentemuseum in Holland which had stipulated the light levels permitted. Those rooms which included delicate drawings, watercolours and oils understandably demanded sensitive lighting design, which would enhance the works whilst protecting them from being exposed to fading through UV. At the same time, the lighting had to avoid abrupt contrasts from the lower levels directed at the drawings to the much higher levels acceptable for the later geometric grids.

Within these technical parameters, the designer, Ray French of SVM Facilities Group, planned a lighting scheme that was essentially cool and calm – not dramatic: this was in keeping with the artist's own formal approach to his work. The exhibition designers, Sean Rainbird and the artist Bridget Riley, also saw the exhibition as having a didactic purpose, and specified white walls and ceilings for a neutral setting. To exploit the reflectance of the white ceilings, purpose-made troughs fitted with uplighting fluorescents were suspended around the perimeter of each room. To accent the art works, adjustable electronic spotlights were used. The fittings chosen also had elongation lenses, which create a lozenge shaped beam rather than a circular one. This is a useful feature for display work, since it allows a more even light distribution than a traditional circular beam.

Lighting at the Mondrian exhibition. Ambient lighting from uplighters in the troughs running parallel to the walls, supplemented by tracked spotlights *(facing)*.

'Stepping out – three centuries of shoes'; 'Art of voodoo'; 'The squeeze chair project': titles of three exhibitions opening around the world in 1998 that show how three-dimensional objects are taking an increasingly important space in museums and exhibitions today.

Not only because contemporary art has moved out of the frame and across the gallery floor, but also because our cultural appreciation is widening. All sorts of objects and artefacts, from traditional craft tools to contemporary commercial packaging, are now within the mainstream exhibition spectrum, while some decades ago they would have been shut up in display cases or kept in study collections.

Faced with the challenge of lighting three-dimensional objects, the lighting designer must try to answer two general questions: what does the visitor need to see, and how does the object demand to be seen? The first question is about the content of the object, and the second about its context. If the object is complex – a clock or scientific instrument, for example – it may need a lot of well-directed light to make its workings clear. Or if there is one aspect to the object that makes it unique or important – Roman lettering under the engraving of a Saxon cross, for example, or an unusual inclusion in a geological specimen – this may need more detailed highlighting.

The imposing galleries of Egyptian sculpture at the British Museum use both natural and artificial light, the latter from fixtures in the ceiling recesses *(facing)*.

If lighting an object from the standpoint of content is relatively objective, selecting a contextual approach may well be more subjective. How an object should be seen refers in part to the object's relative place within the scheme of the exhibition and to the general design approach of the exhibition project, but it can also refer to how the object, if it is a historical one, would have been seen and lit in its original setting. The arguments about lighting historical works for today's audience or lighting them as they would have been seen by their contemporaries move back and forth across the world of museum design and lighting, and there is no definitive answer to this debate. But the original positional aspects of work – for example sculptures originally placed above doorways or along roofs – may need to be considered in devising a lighting answer whatever method is adopted.

Other aspects that will help determine the lighting solution include the scale of the object, not only its intrinsic size but the comparison between the level of detail and the whole; a large detailed bas-relief will require a different approach from a single figure, even a massive one. The texture, colour and reflectance of the object also needs to be considered, and its relationship to other objects on display. And these in turn need to be judged against the proportions of the display area or room available. The other primary consideration, as we have seen, is the susceptibility of the material of the object which needs lighting: this is particularly important for textiles and natural history specimens, but, as has been said, good design practice should always include a conservation check before an object is lit.

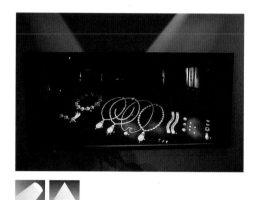

Architectural models by Terry Farrell
exhibited at the RIBA in London. The large
scale of the models meant that fittings could
be included within and below them, to add to
the effects created by the ambient ceiling lights
and tracked spotlighting *(above)*.

Lighting jewellery is a challenging task. The light
needs to be strong enough for detail to be visible,
without glare *(facing, top)*.

African tribal art displayed in a deliberately exotic
way. Note the choice of carpet to bring out colours and
textures in the objects displayed *(facing, below)*.

Bodleian Library,
Oxford, UK.

'Dominus illuminatio mea (God is my light, the motto of Oxford University).'

The Bodleian Library in Oxford is one of the oldest university libraries in Europe, founded by a gift of books from Humphrey, Duke of Gloucester in 1470. As a copyright deposit library it contains copies of all books published in the English language since 1610. It also has major and important collections of foreign publications, rare books, incunabula, manuscripts, engravings, letters and other bibliographical material. In order to present some of these treasures to the public, Bodley's librarian decided to create an exhibition room within the 16th-century building.

The entrance is formed by tall doors panelled in plain wood, with a carved bust of Sir Thomas Bodley, who refounded the library in 1602, in a display cabinet beside them. Within, there is a narrow rectangular room with display cases painted in pale colours with plain wood fittings. The display cases are either vertical boxes, or horizontal desk-type cases with angled, glazed lids. Additional items can be displayed against the walls, hung from a ceiling rod: the walls themselves are hung with pale cloth alternating with plain white panels.

The central quadrangle of the Bodleian Library, Oxford *(facing)*.

The intention of the design is to provide a relaxed and calm setting for the contemplation of small-scale objects. The colours of floors and walls are deliberately neutral. Ambient lighting is provided by wallwashing lights from tracked spots on the longer walls, and single uplighters mounted in the tops of the vertical cases that each throw a soft pool of light on the ceiling. In the vertical cases, fittings are mounted both into the underside of the top and in narrow tracks running down each corner post. Low-voltage medium-beam spotlights have been used: their adaptability means that they can be repositioned for different displays, and the low voltage ensures a minimal heat problem within the cabinets. In the desk-type cabinets similar low-voltage spots are mounted on a track under the top rim.

The exhibition area from the reception desk: note the reading lamp integrated into the reception desktop, and the use of calm but subtle colours throughout *(above)*.

In the vertical cases, glazed shelving allows an even wash of light from the tracked fittings and fixed fittings in the top to fall over the manuscripts and rare books displayed on different levels *(facing)*.

Exhibition Room

While book paper and ink are generally not too light-sensitive, the natural materials (parchment and vellum) and organic inks and paints of illuminated manuscripts are quite sensitive, and all are heat sensitive: the cases are therefore sealed and air-conditioned. The average lux within the display cases is around 100, a combination of the gentle ambient lighting and the accent lighting in the cases.

The entrance doors to the exhibition room. Note the scalloped lighting effects from the wallwashers above the doors. The bust is lit with an overhead compact fluorescent *(top, left)*.

Within the desktop display cases, an even fall of light over each object, whether on the base or the back wall, comes from the tracked spotlights under the upper angle of the top. Ceiling downlighters or an external track would have created reflection problems in the glass lids *(facing)*.

General view of the exhibition room: the design is an invitation to study quietly and closely the work on display. Ambient light from ceiling-tracked spotlights wallwashing the outer walls, uplighters in the display cases, and accent lighting from miniature spotlights in the display cases *(below, left)*.

Earth Galleries, Natural History Museum, London, UK.

'You can't – and shouldn't – iron out the complexity of life.'

Albertopolis is the name given to the complex cluster of museums, concert halls and university buildings created on the site of, and largely with the profits from the Great Exhibition of 1851, patronised and promoted by Prince Albert. Within Albertopolis, Waterhouse's Natural History Museum, with its majestic interior spaces and relentless friezes of animal life, is admired even by those who rightly consider most 19th-century English architecture to be pretentious pastiche without panache. In the 1930s the Geological Museum was built adjacent to it, in the Imperialist Classical style so popular at the time with such diverse national leaders as Adolf Hitler and Joseph Stalin.

The architectural practice Pawson Williams had, in 1993, provided a new setting for the Primates Gallery within Waterhouse's Natural History Museum, a project described by Kenneth Powell as 'one of the few interventions into a great historic museum... which is truly worthy of its setting.' Following the decision to incorporate the Geological Museum into the Natural History Museum, Pawson Williams was again invited to act as architect. In response to the brief, Keith Williams redesigned the existing galleries to create new 'host spaces' – within which exhibition designers would create specific exhibitions – and a new series of entrance spaces, linking the old museum (now called the Earth Galleries) to its new parent. The whole project involved the refurbishment of 6,300 square metres of floor space.

The main staircase leading to the RTZ Atrium, with deliberately diffuse lighting to emphasise the architectural quality of the space *(facing)*.

The new 300-square-metre foyer has been created by radically altering the existing cramped central entrance and extending into the wing spaces on either side. These wing spaces, originally single height and containing storage facilities, have been brought into the main foyer to form a single, double height volume by punching out the existing floor slates and walls above. This move, as well as dramatically increasing the sense of space, allows a sectional transparency to take place, offering glimpses from the foyer to the main atrium level beyond. Fundamentally, the design strategy created space for significant specimens from the Museum's collection to be displayed at a high level within the newly created foyer space.

Detail of the reception area in the Foyer Gallery. The formality of the architectural solution is given sensuous punch by the paleontological exhibits (Megladapis skeleton and Starunia rhinoceros) *(right)*.

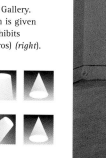

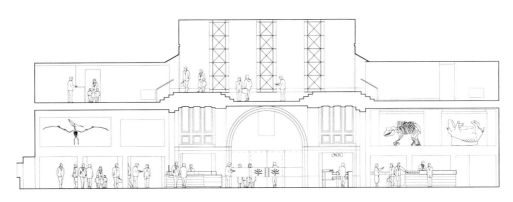

Foyer Gallery

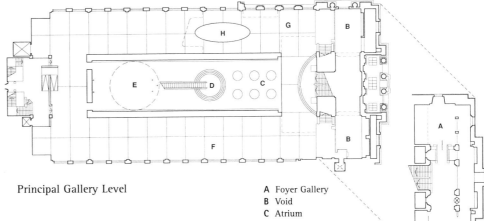

Principal Gallery Level

A Foyer Gallery
B Void
C Atrium
D Escalator
E Rotating Globe Above
F Exhibitions
G Retail
H Function Store

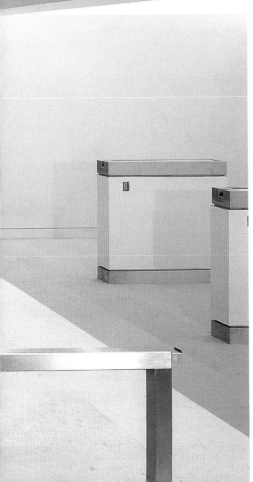

Floor plans of the Earth Galleries showing the entrance level and principal floor. The cross section clearly illustrates the reception area and the display spaces *(above)*.

For the display areas in the Foyer Gallery, the architects looked for a 'white box' approach to create an evenly lit formal background to the work presented *(left)*.

From the point of entry, a spatial sequence has been sculpted, which gradually introduces the visitor through an ascending scale of spatial containment before finally revealing the remodelled main atrium beyond. The best of the existing plaster work and marble has been retained, whilst the new elements have been designed in spare, almost minimal fashion to make clear the new insertions. The palette of new materials was deliberately restricted. Portuguese limestone for the floor, white plastered walls and ceilings, with hot spots of vibrant colour in hard plaster at each end. Contact and meeting areas such as ticket desks and information points are marked with pale American maple and stainless steel, whilst more private areas are shielded from view by full-height acid-etched glass panels.

Pawson Williams defines its approach to architecture in terms of 'contextual minimalism, a reductivist aesthetic combining tradition, abstraction and sensuality.' Lighting plays an important role in its design process to achieve this. At the Earth Galleries the architect's clear design parameters were that the lighting should be discreet, and bounced, reflected or diffused to emphasise the key architectural features of the space. Wallwashers have been positioned over the main staircase entrance. The ticket area is highlighted by downlighters installed in the double height space and the adjacent showcases lit with 50-watt low-voltage lamps in downlighter fittings. Background lighting throughout the remainder of the galleries is provided in a similarly discreet manner.

The scale proportion and quality of light in the new foyer gallery, which is almost totally artificially lit, introduces the visitor in a subtle and restful way to the transformation that has been achieved in the heart of this major international museum.

View looking up the staircase showing downlighters under the ramps *(above)*.

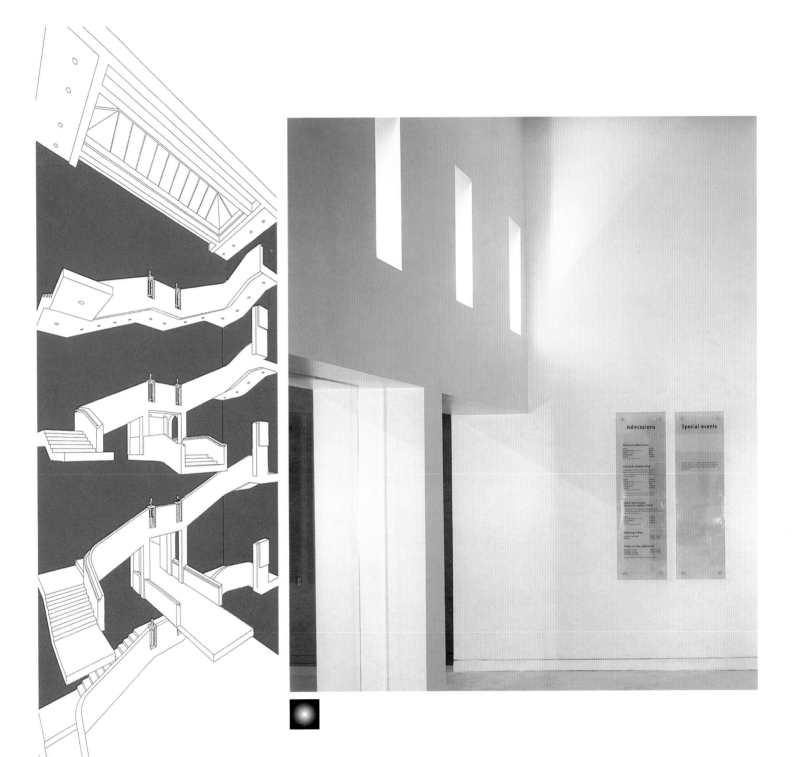

Interior of the entrance space showing diffuse lighting
(above).

Exploded axonometrics of the staircase design *(above, left)*.

The National Gallery of Modern Art, Reykjavik, Iceland.

'Modern work in modern settings.'

The National Gallery of Modern Art in Iceland houses both permanent collections and temporary exhibitions. The formal architecture, crisp and contemporary, makes particular use of rounded windows to supplement artificial light with the clear, cool daylight of this northern latitude. The extent of this challenge, and how it can be met, can be seen in the daytime and night-time views of the same area. In the display area for paintings, the ceiling lights are mounted above a gauze screen to create a diffused wallwashing light.

Lighting bronzes is an exacting business. Too much light and the surface detail disappears into sparkle and reflection, too little and the visual qualities of even the best works are deadened. In such circumstances, the opportunities of a complete system for lighting come into their own. If both fixed overhead lighting and variable sources (for example on track) are available, it is possible to balance the light falling on a sculpture to create the best effect. One which, in addition, needs to be seen from all angles, especially when the sculpture is mounted on a free-standing plinth. Here, working closely with the exhibition organiser will be extremely important.

Bronzes exhibited in the main area showing the benefit of the indirect ceiling lighting and daylight from the round window (facing).

Paintings on display in the main room at
Iceland's National Gallery of Modern Art *(above)*.

The main entrance staircase, with its large glass
rooflight, seen in the daytime *(facing)* and at night
(far, left).

In the smaller exhibition area a range of ceiling
fixtures allow for a more controlled display, despite the
more generous provision of daylight and the highly
reflective wall surfaces *(left)*.

Welsh Contemporary Crafts exhibition, Sana'a, Yemen.

*'*Cymru am byth.*'*

'Best steam coal', in the days of coal-fired naval and merchant ships meant Welsh coal, and the demand for worldwide supplies to fuel the ships of the British Empire led to coaling stations being set up on major trade routes, including at Aden, on the southern end of the Red Sea. From Aden, now in the Republic of Yemen, the local people began to work on the passing ships, and with time the Yemeni community in south Wales became the first Arab community in Britain. By the end of the 19th century, the Cardiff Bay area became an ethnic melting pot, as people from all over the world working in the coal trade settled there, often offering lodgings to seamen and travellers from their own countries. One former resident recalled that each nationality used to put their national flag outside their house, so that arriving sailors could find their fellow countrymen. It made the streets a riot of colours, with all these different communities living side by side. This fact was celebrated in an exhibition of photographs, documents and memorabilia of the Yemeni community shown in Cardiff in 1994, which in turn led to an invitation to present an exhibition with the theme of Contemporary Crafts from Wales in Sana'a in Yemen the following year.

General view of the exhibition area showing the displays laid out on tables with overhead tracked lighting *(facing)*.

Sana'a is the capital of the Yemen, and the largest city after the Red Sea port of Aden, an important coaling station in the 19th century and now a major petroleum bunkering point. The exhibition organiser, Patricia Aithie, was very happy to see the exhibition shown in the Yemen, because it was a useful reminder of cultures meeting each other at an ordinary, human level,

in the past, not as colonial masters and servants, or as migrants or tourists – the settlement in Wales had emerged naturally from the coal trade.

The lighting made all the difference to the exhibition, according to Aithie, not only in improving the appearance of the work on show, but also in making the space itself more

welcoming. As Aithie says, 'the exhibition was a celebration of the past, but also a pledge for the future, as the lighting installation means that the space can now be used in new and different ways by the local community.'

The exhibition area was a simple rectangular room with good natural light. As these different images *(facing, right and above)* show, the adjustable light levels on the track system could either supplement daylight or replace it in the evenings, thus extending the opening hours for the exhibition.

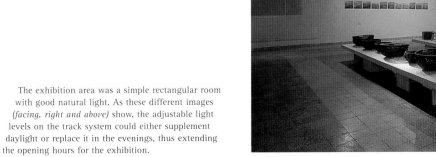

George III Gallery, Science Museum, London, UK.

'Philosophical Tables and orreries, air pumps and armillary spheres.'

The mid-18th century in Britain was not a period of major scientific discoveries, but was a time when popular interest in science grew, through public lectures and aristocratic and royal patronage, especially under King George III, financed scientific expeditions, built an observatory at Kew, near London, and commissioned his own scientific apparatus. One of the popular lecturers of the time was Stephen Demainbray (1710–82) who later worked for King George, and whose collection of instruments, used in his lectures, passed into the royal collection and from there to the Science Museum. Thus the collection includes both grand pieces, made by the best-known instrument makers of the period, George Adams (including his Philosophical Table on which problems in mechanics could be displayed), and simpler instruments used in demonstrations and lectures, often made by the lecturers themselves, such as Demainbray's own model of Valoué's pile-driver.

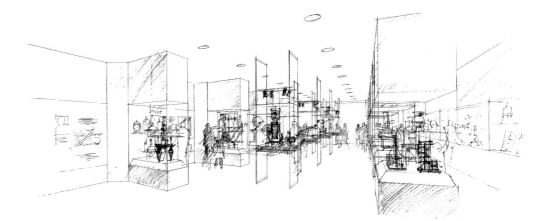

General view of the George III room *(facing)*.

Alan Irvine's sketch view of the installation *(right)*.

The decision in 1991 to create a new gallery for the collection was the first stage in a general plan to refurbish the Science Museum. The designer Alan Irvine was asked to design the gallery on the third floor. He proposed a wall of informational texts, diagrams and illustrations, with an audio-visual display, and a series of display cases, some free-standing and some wall-mounted to show the objects in the collection.

As part of the design process, the museum's conservation experts considered the acceptable light levels and the necessary technical solutions. Their report looks at the lighting in the four different types of display cases (A, B, C & W), as well as the ambient light level. Included here is an extract from the report as an example of the kind of technical expertise alongside which the lighting designer has to work. It is dated September 1993.

Entrance to the George III Gallery. Wallwashing from ceiling fittings highlight descriptive texts *(above, left)*.

George III Gallery Conservation Report

1. General Light Levels

Extremely light-sensitive material must not be exposed to light intensity of more than 50 lux. All other material must not be exposed to light intensity of more than 200 lux. Ultraviolet levels must be not more than 75W per lumen. These are currently accepted figures.

The quoted maximum allowable exposure levels per annum are as follows, based on 50 hours of lighting per week, and 43 weeks to a museum year:

General material (lit at 200 lux): 650 kilolux hrs/yr.

Highly light-sensitive material (lit at 50 lux): 200 kilolux hrs/yr.

These figures are based on an eight-hour day for six days per week. Since the Science Museum is open for eight hours per day, for seven days a week, closing only for an average of three days per year, the light levels in the gallery must be kept to or below the 200 and 50 lux levels as above.

After the case lighting has been reduced to acceptable levels, the ambient lighting can be sufficiently reduced to highlight the cases in the gallery.

2. Gallery Light Levels

Light levels in the gallery were assessed on the 9th September with the help of the Electrical Services Manager. Light intensity (lux) and ultraviolet (UV) levels were read:

Maximum ambient gallery lighting: 390 lux at three feet from the floor.

UV levels negligible.

The ambient lighting can be reduced since the system is on a dimmer switch.

D Cases: 12–112 lux at floor level.

UV levels negligible.

Lighting consists of spotlights only.

Current levels in the D cases are acceptable.

B Cases: 1100 lux on case base. 100W per lumen (UV).

Four lights on exterior of glass case top.

A, C and W Cases: These are lit with two systems; spots and fluorescent tubes.

UV levels are less than 75W per lumen.

Spotlights: < 450 lux

Fluorescent tubes:

A Cases: 800 lux on case base to > 2000 lux at case top.

Light fittings consist of: two sets of Twin tubes approximately five feet in length (i.e. four tubes).

C Cases: Approximately 2000 lux on case base.

Light fittings consist of: two Twin tubes (as for A Cases).

W Cases: 725 lux on case base.

Light fittings consist of: one set of Twin tubes.

3. Conclusion

The light intensity levels in the gallery need to be reduced to the above-mentioned levels. Lux levels within cases will be lowest on the case bases, increasing with height towards the light fittings. Assuming that the use of the fluorescent lighting will be required, removal of one fluorescent tube from each twin fitting will reduce the current lux levels to approximately 1000 lux, or half the current levels. This level must be further reduced using the appropriate neutral density filters, available from Sun-X. The film has been designed for museum purposes, is relatively inexpensive, and will not affect the appearance of the objects or cases. The film is available as a flat sheet which will be laid on the base of the inside of the light boxes in Cases A, C and W. The filters will require special application to the outside of the glass tops of the B Cases by the Company, and if carried out correctly in this way, should not interfere with their appearance.

I will liaise with the Electrical Services Manager regarding the removal of the fluorescent tubes. I can arrange to place the remaining film in the light boxes. I would like application to the B Cases to be carried out as swiftly as possible, given that case dressing starts on the 4th October. Light levels in each individual case and for specific objects will be measured by conservation thereafter.

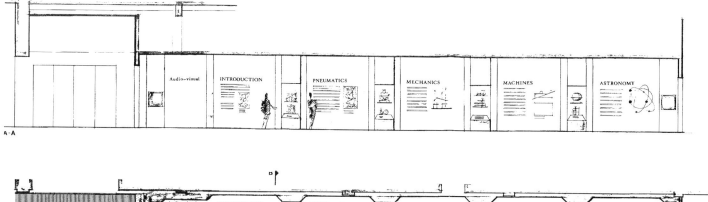

A-A

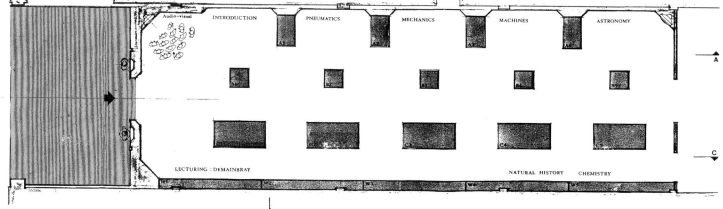

The final solution is a calm and ordered space, with the objects clearly presented and well lit. The teamwork between the designers, lighting experts, and conservation, curatorial and educational staff at the museum has created this success: just as George III worked together with James Cook to observe the transit of Venus across the sun in 1769, the one from his boat in the South Seas, His Majesty from Kew.

Alan Irvine's first-stage plan and section for the gallery *(above)*.

Display cases and wall-mounted information panels make up the contents of the gallery. Note that the complex structures of the instruments, in particular, require carefully angled lighting to be clearly visible, while natural materials (such as leather cases and mounts) need controlled lighting. Balancing these two elements requires fine judgement *(facing)*.

Burrell Collection, Glasgow, Scotland.

'In all history, no municipality has ever received from one of its native sons a gift of such munificence.'

The Burrell Collection, given to the city of Glasgow in Scotland in 1944, was a collection in the grand tradition: paintings, sculptures, architectural antiquities, porcelain and lacquer from across the centuries and across the world. The collection both reflected the tastes of its creator, the shipping magnate Sir William Burrell, and Glasgow's tradition as a great port and mercantile centre. The decision to house the collection in a new building just outside the city centre at Pollok Park was an exciting opportunity for the architect Barry Gasson, who won the two-stage architectural competition to create it.

'The design grew from some basic ideas,' he explained in a book published to celebrate the opening of the collection in its new home. 'The first, and the most important, was that this was to be a collection in a park, not in a city. This offered the opportunity of making the grass, the trees, the woodland plants, the bluebells and bracken, a context for the display of the collection. The second consideration was to resolve the problems posed by the elements of the collection: how to incorporate the three Hutton Rooms, the stone arches and windows, the timber screens and ceiling, the tapestries and stained glass, the paintings, carpets and many objects of stone, metal and ceramic.' (The Hutton Rooms were facsimile reconstructions of three rooms at Hutton Castle, Sir William Burrell's home until his death in 1958 at the age of 96: they are the only specific memorial to him, as there is no portrait of him in the collection, at his express request.)

View within a view at the Burrell Collection, Pollok Park, Glasgow *(facing)*.

The building housing the collection, in plan a truncated right-angled triangle, was placed at the top of a sloping park, backing on to woodland on the hypotenuse and the cut end. It is a single-storey structure with a mezzanine. The Hutton Rooms frame the interior entrance courtyard, with other rooms showing light-sensitive exhibits, such as tapestry, within the main body of the building. But around the edges, whether opening on to woods or parkland, daylight from both windows and ceilings has been deliberately introduced. This is supplemented by track lighting along the ceiling edges. It was the architect's concept to make a visit to the collection a walk in the woods, and so the daylight element is essential. The lighting designer's task was to supplement this where necessary, and to provide alternative lighting in spaces where daylight was inappropriate or inaccessible.

One particular feature of the collection is the 20 or so architectural antiquities – doorways and portals, windows and niches, mainly from mediaeval Europe but also from other cultures. (Some of these were acquired by Sir William, late in his life, from another famous collection, that of the American newspaper tycoon Randolph William Hearst.) Where possible, these items were used for their proper functions: the 16th-century sandstone doorway from Hornby Castle in Yorkshire is the entrance to one of the sculpture courts, and a linenfold-panelled wooden door-frame gives access to the costume and needlework collection. Here supplementary lighting from tracks is used to create appropriate levels and conditions of light, dappling the sandstone as sunlight would have when the door stood in the open, or highlighting the intricate decoration of a 12th-century limestone portal from France.

Lord Norwich describes the objects that comprise the Burrell Collection as leaving one spellbound – not only for their quality, but also for their quantity and the quite astonishing breadth of taste they represent, and he goes on to describe their new home as surely the most important new museum to have been built in Britain during the last century. The challenge of presenting this array of riches in the right light was achieved through partnership between the architect, the curators and the lighting designers. It is a great success, particularly given the important role given to daylight as a presentation medium in this museum in a park.

The important tapestries in the collection were exhibited in the centre of the building, to minimise exposure to daylight, and lit under tightly controlled conditions for UV radiation *(facing and right)*.

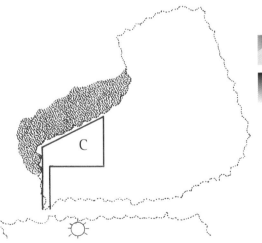

In the sculpture collection, display cases are lit by ambient daylight and spotlights on ceiling tracks *(above)*.

Architectural antiquities on a large scale are one of the collections specialities, and doorways in particular, which have been used as proper entrances wherever possible. The different textures, colours and depth of carving on each one required special handling, though all are lit from the main track system *(facing)*.

The general plan shows the Collection Building C at the top of open grassland, with woodland behind and beside it *(above, right)*.

(far

Each space to be lit, interior or exterior, presents a different challenge, even if the approach can often be similar and the values sought in the final result also much the same. Part of the fun of lighting design, after all, is that every job is different. As far as spaces for public display are concerned, the definitions of exhibitions are extending all the time, as we have seen: so are definitions of culture.

Our desire to experience and share as wide a range of events or things outside our immediate surroundings seems to be expanding, and technology and imagination – through the media wall and the interactive touch screen, through simulations and life-size animated models, through programmable 'rides' and historic reconstructions – are coming forward to deliver new experiences all the time.

This section celebrates a few of the best, and least usual, recent solutions to the challenges of experiential spaces: at Carrickfergus, the recreation of chivalry; at the Cartier Foundation, the excitement of a glass-walled exhibition area; in Birmingham, maintaining the values of an existing building while giving it a new purpose; at the Design Museum, using false walls to create special lighting effects. Such opportunities for lighting creativity do not occur every day, so these are not put forward as blueprints, but as clues to the possibilities of designing with light.

Birmingham School of Arts, Birmingham, UK.

When the Birmingham Municipal School of Arts and Crafts, designed by Victorian Birmingham's leading architects Henry Chamberlain and William Martin, opened in 1885, it was the world's first purpose-built municipal school of art and its students achieved the highest standards in the country. Not surprisingly, it was a source of great civic pride for industrial Birmingham. That pride was revived when in October 1995, after two years of work, Birmingham School of Art reopened for the start of the academic year, offering its staff and students a grand Gothic revival building that had been completely refurbished and in part rebuilt, fit for continued use into the next century.

'Applying a contemporary aesthetic to custom designing a fitting for a 19th-century building.'

Over the 50 years since the Second World War, the buildings had declined through lack of maintenance, *ad hoc* additional servicing and a lack of appreciation for the quality of the building: the fine brick and stone surfaces of the museum hall had been obscured with white paint. The refurbishment also involved creating new elements within the building such as the mezzanine floors, which were designed in a wholly contemporary manner, making clear the distinction between the new and existing elements of the building.

The client and the architects discussed applying this aesthetic to the lighting with the Specials Department at Concord, who suggested that a custom fitting could be provided, custom designed and made to suit the building. After a series of sketches and mock-ups, the profile was agreed upon and the total system was then technically developed and manufactured.

The upper hall of the Birmingham School of Art with its cross-beam roof and encaustic tile floor. Daylight is supplemented by track fittings beneath the angled skylights in the roof *(facing)*.

The lighting now consists of a series of bays, defined by a very flexible linear system housing louvered fluorescents providing low brightness or lighting to Category 2 lamps using T8 fluorescent lamps in 36-watt, 58-watt and 70-watt modules. The system has facilities for uplighting, as well as downlighting and can carry track for spotlights. The new profile and size visually complements the interior spaces: even the louvres have been curved to complete the design detail. The system is constructed with aluminium extrusions and a facility for continuous wiring, and has also been wire-suspended from the ceiling in some areas, while in others, it has been wall-mounted.

According to Matthew Goer of Associated Architects: 'The opportunity to promote bespoke lighting for the project was particularly welcome. We were able to design a system appropriate for the large studio spaces with confidence that it would be developed into a thoroughly resolved finished product. The lighting has now been installed and works well aesthetically and functionally; the quality of light in the refurbished building is now superb.'

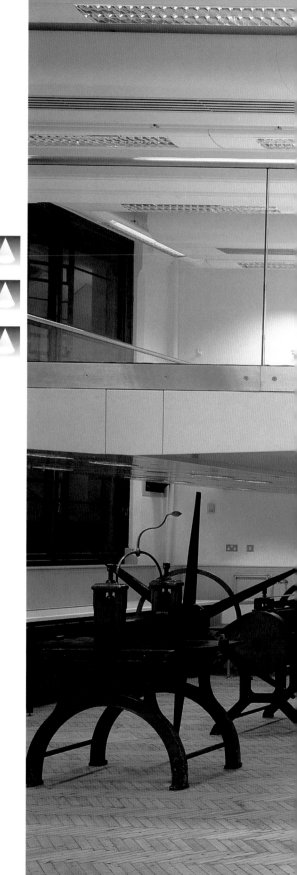

Split levels were inserted to increase working space. The top area is lit by combined fluorescent uplighters and downlighters, the lower by recessed compact fluorescent downlighters in the new ceiling *(right)*.

The main areas were lit by suspended fluorescent downlighters on specially designed battens. Suspension was used to interfere minimally with the existing structure *(left)*. Note that the battens are also fitted with spotlights on each end for additional detail lighting. These can either be placed above the fitting *(left)* to light architectural elements, or below it *(above)* to add light to wall surfaces.

Benetton Sport '90, The Design Museum, London, UK.

'United Colors
of Sport and Light.'

Design in museums is not new: the Victoria and Albert Museum in London was created at the end of the 19th century as a treasure house of the best craft and industrial production, as well as holding paintings and drawings. New York's Museum of Modern Art put the acquisition of contemporary design on its agenda almost from its foundation in the early 1930s. But the Design Museum in London, created by Sir Terence Conran and Stephen Bayley, was among the first purpose-built museums of design in the world. It both houses a permanent teaching collection of 20th century industrial design objects and hosts temporary exhibitions on contemporary design.

The first floor of the museum is the main exhibition area, an enclosed but uncluttered space with polished concrete floor and white walls. While there is a track lighting system available, the exhibition space is treated very much as a shell, into which a predesigned or specially designed exhibition can be fitted: often Design Museum exhibitions are as interesting for their expo design as for their content!

The main exhibition, showing information banners and display stands lit from narrow spots on ceiling-mounted tracks *(facing)*.

One such exhibition was Sport '90, sponsored by the Italian fashion manufacturers Benetton in the early 1990s and focusing on new designs for sports equipment and clothing. Because all the material on show was new, conservation problems were minimal, and the lighting designers were able to work alongside the exhibition team to create a new and special environment. The objects were exhibited on large raised platforms with partial walls. Lighting was installed under the platforms and behind the walls, so as to create bands of light spilling out of the gaps. Coloured light was used here, while the objects themselves were lit by narrow beam spotlights from the overhead track from points on the top of the false walls.

The coloured lighting had a purpose in linking groups of objects that shared a common theme, for example the use of new materials in sports goods, while the hard white light from the spots clearly identified and configured the individual objects on display. The whole invited the visitor to understand the new role design was playing in the world of sport, traditionally dominated by craft practices and traditions.

General views of the exhibition showing how lighting behind false walls and under the raised floor added bars of coloured light, marking each context, to the detail lighting from ceiling spotlights *(above and facing)*.

Carrickfergus Heritage Centre, Carrickfergus, Co. Antrim, Ireland.

'An exciting, eventful environment containing history and entertainment in roughly equal measure.'

Heritage is an increasingly important concept in the development of national identity, culture and tourism. In identity, places and artefacts – the giant redwood forests of North America, the palace of Versailles, the Great Wall of China, the paintings of Constable or Vermeer – are seen as part of the definition of a country or people. In culture, similar things are seen as visual history, a facet of the qualities of a society. And in tourism, museums and historic sites attract and explain a city, county or country to visitors. But a great deal of history only exists in the written record. To what extent is it permissible or useful to recreate an historic event or rebuild an historic artefact? Reproductions of early sailing vessels, such as the *Kon-Tiki*, have taught us much about early navigation, and recreated early industrial sites such as Telford have considerable educational value. But pastiche can become parody, and so it may be better to offer historical information in a wholly enjoyable way, in a context and using techniques that are wholly modern, so that the recreated does not get confused with the real.

Carrickfergus is a town and port on the north shore of Belfast Lough in Northern Ireland. Above it, Carrickfergus Castle, built in 1180 by the Normans to control the unruly Irish, has looked down on 800 years of turbulent Irish history, and still stands. And in the town below, what was once a depot for disposal trucks has been transformed into the new 'Knight Ride Centre'. Here specially designed cars carry visitors through tableaux representing 1,000 years of Carrickfergus' stormy history, brought to life by sound, figures, audio-visual and stage effects, together with lighting.

In the lower-level sets and scenes, generally lower levels of indirect lighting were used *(facing)*.

Track systems are fixed to the support beams of the glazed roof, providing four circuits at high level so that special effects can be created by dimming and switching. Display hooks fitted into the track can carry banners or other suspended exhibits. During daylight hours, shafts of light beam down on to the displays from powerful 150-watt single-ended metal halide lamps in spotlight fittings whose tight beam reflectors provide a strong dramatic accent. 3000 Kelvin lamps bathe the exhibits in warm light. Low-voltage projectors with gobo holders and colour filters can easily handle the requirements of changing the visual graphics projected into delicate fabric banners and panels. These are linked to programmed switching units to trigger the display effects. At night, daylight is simulated by the 70-watt metal halide floodlights washing the south side of the glazed atrium with light, thus highlighting the beam structure whilst suggesting a directional wash of light through the glazing.

Architect Stephen Wyatt has transformed the original warehouse-type building, by replacing the central section with a double-glazed curtain wall and a glass roof which spans the original – but newly cleaned – red-brick wings. The entrance is guarded by a metal sculpture of three sword-wielding knights; above them hang banners with Celtic artwork and lettering. Within the space, the balconies that ring the upper level are supported by welded cylindrical columns. On the ground floor, there are seven shop units and a raised dining area at the end of the mall.

In the main exhibition area, visitors take a tour in specially designed cars, through a series of tableaux depicting key events in the history of Antrim. The lighting scheme, created closely with the architect, sympathetically enhances the multi-purpose space. Using the latest energy-efficient technology, the scheme complements the architectural detail while supplying unobtrusive lighting with optimum flexibility. The system copes with the changes from day to night, altering visual effects on localised displays and also the need to supply decorative luminaires to maintain the architectural theme.

The walls and walkways around the perimeter of the space are lit by special decorative lanterns, designed to complement the architectural aesthetic. They are fitted with compact fluorescent globe lamps which have a life of around 8000 hours and a lumen output similar to

a 100-watt GLS lamp. By having bright walls, the whole space appears bright although overall lighting levels can be reduced to no more than 200 lux. The sculptures guarding the exterior entrance are lit with a buried floodlight washing the rock podium with an eerie cold blue light. In deliberate contrast, the faces of the knights are given warmth and colour by two light-beam SON floodlights shining up from the base of the sculpture.

The total effect at Carrickfergus is much nearer to the conventions of theatre than the traditional world of museums, and not surprisingly the lighting response needed to use the kind of controls and effects more commonly found on stage. The result is an exciting, eventful environment, containing history and entertainment in roughly equal measure.

The general lighting principle was to flood the ceiling with light, picking out details such as the sail-like floating banners and the imitation fish *(above)*.

Metal halide spotlights with barn-door fittings were used to direct wide swathes of light on to specific areas *(facing)*.

JEAN MICHEL ALBEROLA

DU 3 MARS AU 16 AVRIL 1995

Cartier Foundation,
Paris, France.

'Creating unreality through light and reflection.'

The concept of an exhibition area completely walled in glass may seem deliberately obtuse, but for the French architect Jean Nouvel's Cartier building in Paris it was the answer to a specific challenge placed by the site, and an opportunity to examine aspects of his architectural ideals. The site is on the Boulevard Raspail in southern Paris and contains a tree of liberty planted by the writer Chateaubriand 150 years ago. Previously the site had been the home of the American Center, a small building set behind the tree, surrounded by a garden and separated from the boulevard by a stone wall. These two features, the tree and the garden, were to be retained in the new project.

Nouvel has also been experimenting extensively with the concept of dematerialisation in architecture – the extent to which a building can achieve visual anonymity or insubstantiality – for example, in his Cologne media tower project the walls were to consist of immense electronic displays, the building itself becoming a source of information for the surrounding city. At the Cartier building, the opportunity to test this concept in a new way arose. By using large glass panels on an extremely thin steel skeleton, Nouvel was able to create a building that was virtually transparent, thus allowing the famous tree and the garden to remain visible. The stone wall at the front of the site was replaced by a five-metre-high transparent glass wall, and this and the parallel glass planes of the building behind allow a series of visual reflections and reactions to light to run across and through the whole space.

Exhibition installation by Jean-Michel Alberola in the main gallery *(facing)*.

The Cartier building under different daylight conditions photographed by Philippe Ruault: as Nouvel himself says, 'Sometimes even I wonder whether I'm seeing the building's transparency or its reflection!' *(following pages)*.

The exhibition area is on the ground floor, with a small studio theatre in the basement below. The offices for the administration of the centre, and Cartier's own office space are in the central tower. Here too the principle of muted transparency is maintained with ground glass screens providing privacy for office areas, and narrow grey tables and filing cabinets designed by Nouvel specifically for the building. The main exhibition area is a double-height cube projecting from the side of the main block, and lit from the recessed metal halide fixtures in the dark ceiling. These have special dimming arrangements to maintain an even lighting level throughout the day and evening when the centre is open. Despite the transparent walls, the exhibition area works remarkably well, in part because of a lighting system that concentrates the visitor's attention on the work on display.

Installation by the lighting designer Ingo Maurer, who planted several thousand small luminous coloured tubes in the garden and gallery *(right)*.

'Lord Finchley tried to mend the Electric Light

Himself. It struck him dead: and serve him right!

It is the business of the wealthy man

To give employment to the artisan.'

While we might not share Hilaire Belloc's view of the social order, he was quite right to insist on bringing in the professional to do the job! Where historic buildings or valuable works of art are concerned, this is doubly important. I have tried to show here how important it is for the lighting designer to be an integral part of the team creating the total ambience of a public space, a permanent display or a temporary exhibition. What I hope is also clear is the excitement that such an invitation offers for the creative use of light in communicating the quality of an environment, or the visual drama of works of art, contemporary or ancient, to a wider public. This challenge is what has made lighting exhibitions and public spaces one of the most enjoyable aspects of my career as a designer.

In the public rooms and spaces of Jean Nouvel's Institut du Monde Arabe in Paris, white light shimmers through a wall of alabaster or is muted by subtly opening and closing lenses. In the nave of St Paul's Cathedral in London the gentle play of light reveals the glory of Wren's architectural achievement. In Frank Lloyd Wright's Guggenheim Museum in New York the works on display lift from the walls around the spiral ramp. And in the immense hall of Renzo Piano's Kansai airport the detail of the kilometre-long space can be appreciated both by night and by day.

In all such public spaces, the lighting designer has a series of challenges to meet. Today there is an immense array of lamps, fittings and control gear available for this task. With the increasing importance of art exhibitions and museums in our understanding of the cultures of the world, in the celebration of the city centre and the public forum as a social space, and with the realisation that spaces must communicate with their public through their architecture and design, the opportunities for creative lighting are also increasing.

At the heart of successful lighting design lies a triumvirate of basic concepts. The first of these is knowledge. The lighting designer today has to be aware of the technology available, in lamps, fittings and control gear, and of the designer's obligations towards technology and the environment – to be conscious of energy use, and seek to limit it, to be careful over glare and wastage of light, to select products that are safe. The second is communication, the ability to exchange ideas and concepts with the team working on the task, to understand the aims and objects of the brief, not just in its written form but in terms of the aspirations and intentions of the client and the public user. Communication is at the heart of teamwork. And the third quality is imagination and understanding, an awareness of how a future space is going to look, how it is going to be used, how it is going to be enjoyed. Making a space better through lighting – more pleasant to walk through, easier to learn from, happier to be in – is one of the best achievements a lighting designer can hope for.

Photograph© Richard Davies

Exhibition installation by the architectural
group Future Systems at the Institute of
Contemporary Arts in London *(facing, and right)*.

GLOSSARY

Adaptation: The process which takes place as the visual system adjusts itself to the brightness or the colour (chromatic adaptation) of the visual field. The term is also used, usually qualified, to denote the final stage of this process.

Apparent colour: Of a light source; subjectively the hue of the source or of a white surface illuminated by the source; the degree of warmth associated with the source colour. Lamps of low correlated colour temperatures are usually described as having a warm apparent colour, and lamps of high correlated colour temperature as having a cold apparent colour.

Average illuminance (Eave): The arithmetic mean illuminance over the specified surface.

Brightness: The subjective response to luminance in the field of view dependent upon the adaptation of the eye.

Candela (cd): The SI unit of luminous intensity, equal to one lumen per steradian.

Chroma: In the Munsell system, an index of saturation of colour ranging from 0 for neutral grey to ten or over for strong colours. A low chroma implies a pastel shade.

Colour constancy: The condition resulting from the process of chromatic adaptation whereby the colour of objects is not perceived to change greatly under a wide range of lighting conditions both in terms of colour quality and luminance.

Colour rendering: A general expression for the appearance of surface colours when illuminated by light from a given source compared, consciously or unconsciously, with their appearance under light from some reference source. Good colour rendering implies similarity of appearance to that under an acceptable light source, such as daylight. Typical areas requiring good or excellent colour rendering are quality control areas and laboratories where colour evaluation takes place.

Colour temperature (Tc, unit: K): The temperature of a full radiator which emits radiation of the same chromaticity as the radiator being considered.

Contrast: A term that is used subjectively and objectively. Subjectively it describes the difference in appearance of two parts of a visual field seen simultaneously or successively. The difference may be one of brightness or colour, or both. Objectively, the term expresses the luminance difference between the two parts of the field.

Diffuse reflection: Reflection in which the reflected light is diffused and there is no significant specular reflection, as from a matt paint.

Directional lighting: Lighting designed to illuminate a task or surface predominantly from one direction.

Discharge lamp: A lamp in which the light is produced either directly or by the excitation of phosphors by an electric discharge through a gas, a metal vapour or a mixture of several gases and vapours.

Downlighter: Direct lighting luminaires from which light is emitted only within relatively small angles to the downward vertical.

Fluorescent lamp: This category of lamps functions by converting ultraviolet energy (created by an electrical discharge in mercury vapour) into visible light through interaction with the phosphor coating of the tube.

Glare: The discomfort or impairment of vision experienced when parts of the visual field are excessively bright in relation to the general surroundings.

Hue: Colour in the sense of red, or yellow or green etc. (See also Munsell.)

Illuminance (E, units: lm/m2, lux): The luminous flux density at a surface, i.e. the luminous flux incident per unit area. This quantity was formerly known as the illumination value or illumination level.

Incandescent lamp: A lamp in which light is produced by a filament heated to incandescence by the passage of an electric current.

Lumen (lm): The SI unit of luminous flux, used in describing a quantity of light emitted by a source or received by a surface. A small source which has a uniform luminous intensity of one candela emits a total of 4 x pi lumens in all directions and emits one lumen within a unit solid angle, i.e. 1 steradian.

Luminance (L, unit: cd/m2): The physical measure of the stimulus which produces the sensation of brightness measured by the luminous intensity of the light emitted or reflected in a given direction from a surface element, divided by the projected area of the element in the same direction. The SI unit of luminance is the candela per square metre.

Lux (lux): The SI unit of illuminance, equal to one lumen per square metre (lm/m2).

Munsell system: A system of surface colour classification using uniform colour scales of hue, value and chroma. A typical Munsell designation of a colour is 7.5 BG6/2, where 7.5 BG (blue green) is the hue reference, 6 is the value and 2 is the chroma reference number.

Optical radiation: That part of the electromagnetic spectrum from 100nm to 1mm.

Purity: A measure of the proportions of the amounts of the monochromatic and specified achromatic light stimuli that, when additively mixed, match the colour stimulus. The proportions can be measured in different ways yielding either colorimetric purity or excitation purity.

Reflectance (factor) (R, p): The ratio of the luminous flux reflected from a surface to the luminous flux incident on it. Except for matt surfaces, reflectance depends on how the surface is illuminated but especially on the direction of the incident light and its spectral distribution. The value is always less than unity and is expressed as either a decimal or as a percentage.

Saturation: The subjective estimate of the amount of pure chromatic colour present in a sample, judged in proportion to its brightness.

Uplighter: Luminaires which direct most of the light upwards on to the ceiling or upper walls in order to illuminate the working plane by means of reflection.

Utilisation factor (UF): The proportion of the luminous flux emitted by the lamps which reaches the working plane.

Value: In the Munsell system, an index of the lightness of a surface ranging from 0 (black) to 10 (white). Approximately related to percentage reflectance by the relationship $R = V(V-1)$ where R is reflectance (%) and V is value.

Working plane: The horizontal, vertical, or inclined plane in which the visual task lies. If no information is available, the working plane may be considered to be horizontal and at 0.8m above the floor.

FURTHER INFORMATION

As well as specialised trade journals, the major architectural magazines carry regular articles and supplements on lighting design, both products and projects.

The major manufacturers of lamps and fittings publish extensive catalogues, which often contain technical supplements and examples of recent lighting projects. Many of these manufacturers have their own websites, which are useful sources for the most recent information.

Lighting industry trade groups and federations also often publish lighting guides, often with information on local requirements for lighting levels. For specific regulations on lighting levels in public buildings in particular cities, states or countries, the local or national planning regulations should be consulted.

Specific advice on lighting works of art should be sought, as we have often seen in this book, from the conservation departments of museums, or from the National Museum Association.

Major trade fairs for lighting are held every April in Hanover, Germany, and every two years, also in April, as part of the Salone del Mobile in Milan, Italy. These are an excellent opportunity to see a range of new lamps and fittings.

Recent books which deal with general issues of lighting include:

C. Gardner & B. Hannaford, *Lighting Design* (Design Council, 1993)

T. Porter, *The Architect's Eye* (International Thompson, 1997)

J. Turner, *Lighting*: *An Introduction to Light, Lighting and Light Use* (Batsford, 1994)

J. Turner, DESIGNING WITH LIGHT: *Retail Spaces: Lighting Solutions for Shops, Malls and Markets* (RotoVision SA, 1998)

Index

Aithie, Patricia 116
Art & Power exhibition 16
Artificial light 30–35
Associated Architects 133, 134
Atlantis Gallery 80–83
Behrman, Sidney 85
Bellinzona 54
Belloc, Hilaire 153
Benetton Sport '90 138–141
Birmingham School of Arts 132–137
Bodleian Library 96–101
Bodley, Sir Thomas 97
Borne, Ludwig 71
British Museum, London 92
Burrell Collection 124–129
Burrell, Sir William 125
Cardiff 115
Carrickfergus Heritage Centre 142–145
Cartier Foundation, Paris 146–151
Chinese art 11
CNRS 60
Colour 24–25, 28–29
Computer programs 41–42
Conservation, role of 10, 44–49, 121
Cook, James 122
CZWG 85
De Buyster, Guérine 61
Dempsey, Andrew 16
Design Museum, London 138–141
Diffusion 41
Downie, Craig 65
Duveen, Joseph 85
Earth Galleries 102–109
EDF 59

Farrell, Terry 55, 95
Fitting types 36–37
Future Systems 154–155
Gasson, Barry 125
Gehry, Frank 57
George III 119, 122
Genius of Venice exhibition 72
Glasgow 125
Glassware 48
Goer, Matthew 134
Goodwood Visitor's Centre 64–69
Gough, Piers 85
Goujon, Jean 60, 61
Gravity & Grace exhibition 15
Guggenheim Museum, Bilbao 57, 73
Hayward Gallery, 8–17
Intensity 41
Irvine, Alan 72, 73, 119, 122
Lamp types 32–35
Le Vau 59
Light distribution 39, 40, 43
Light pollution 56
Light sources 26–27
Lighting angle 74
Lighting as a system 50–51
Lighting audit 10
Louvre Museum, London 46–47, 58–63, 73
Lutyens, Sir Edwin 10
Maurer, Ingo 150–151
Menil Gallery, USA 73
Mensi Corporation 60
Mexican art 17
Moitte, Jean Guillaume 62
Mondrian exhibition 90–91

Munsell system 24–25
National Gallery, Berlin 79
National Gallery of Modern Art, Reykjavik 110–113
National Gallery, Manila 78
National Portrait Gallery, London 84–89
Natural History Museum, London 102–109
Newton Institute, Cambridge 57
Newton, Sir Isaac 19–21
Nouvel, Jean 147, 153
Output 38
Pawson Williams Architects 102–109
Pei, I.M. 59
Perception 23
Phostére 60
Poe, Edgar Allan 71
Pompidou Centre, France 73
Rainbird, Sean 91
Reflectance 29
Riley, Bridget 91
Royal Academy, London 73
Royal National Theatre, London 50, 51
Ruault, Philippe 147–149
Science Museum, London 118–123
St Katharine's Dock, London 54–55
Stanton Williams Architects 50–51
Tate Gallery, London 27, 77, 90–91
Twilight of the Tsars exhibition 12, 13
Van der Rohe, Mies 79
Visible light 22
Warhol, Andy 76, 77
Welsh Contemporary Crafts exhibition 114–117
Wright, Joseph 31
Yemen Republic 115
Yves Klein exhibition 17